Chemistry Laboratory Manual

George W. Robinson

Emeritus Professor of Chemistry
Southern Polytechnic State University

Second Edition

KENDALL/HUNT PUBLISHING COMPANY
4050 Westmark Drive   Dubuque, Iowa 52002

Copyright © 1992, 2003 by George W. Robinson

ISBN 0-7575-0384-5

Printed in the United States of America
10 9 8 7 6 5 4 3 2 1

Table of Contents

Purpose of the Laboratory

Chemistry is an experimental science, and you cannot expect to gain any real understanding of chemistry without developing some first–hand feeling for laboratory procedure. During this laboratory experience, you will examine properties of various chemicals and perform experiments in which you collect data and see how chemical knowledge is obtained. Hopefully after this experience, you will understand some methods of scientific investigation and will have acquired a firm grasp of the chemical principles discussed in class.

Instructions for Laboratory Work

1. Read and know the safety and laboratory rules and obey them. They help safeguard the interests of all.

2. Read the assignment *before* coming into the laboratory.

3. Bring your laboratory manual, textbook, pencil and calculator to each laboratory.

4. Enter data onto the laboratory reports as you perform the experiment. In this laboratory, you may work in pencil.

5. The laboratory report is usually due at the end of each laboratory session. Your instructor will indicate if the report may be submitted at another time.

6. Talk over the experiments with your classmates. Scientists and technologists learn much by discussion with one another. In many laboratories, you will work as part of a team.

7. Your instructor will indicate the criteria for grading your laboratory reports.

8. Cultivate the habits of curiosity and observation. Do not mechanically follow the instructions like a recipe. Remember that common sense knowledge also applies to the laboratory. *Think!!!*

Safety Rules for the Laboratory

1. You must wear approved safety eye wear at all times in the laboratory.

2. Note and record the location of essential items of safety equipment:

   a. fire extinguishers _____

   b. eye wash stations _____

   c. safety shower _____

3. Note the location of exits from the laboratory.

4. No smoking, eating or drinking is allowed in the laboratory.

5. ***Common sense*** is the basic safety rule of all chemistry laboratory work.

6. Be careful with the glassware, and never use cracked or broken glassware. Discard broken glass in the special container available in the laboratory.

7. When using a flame, make certain that no flammable materials (including long hair) are nearby. When heating a test tube, never point the tube at anyone or look directly into it. The contents can boil rapidly and "bump" out of the tube.

8. Treat all chemicals as if they were toxic and hazardous. Although you will not be working with any extremely dangerous substances, never taste a chemical, wash any chemicals off your hands with soap and water, and immediately clean up any spills on the counter tops. Avoid rubbing your eyes while working in the laboratory. If you are instructed to smell a chemical, fan the vapors toward your nose and sniff gently. Be careful to use only the specified amounts of any chemical. Also get in the habit of washing your hands shortly after leaving the laboratory for the day.

9. Discard excess chemicals and waste products in the proper containers as directed by your laboratory instructor. Never pour solutions down the sink without permission.

10. Because cleanliness and safety are related, keep your work area in order. Also keep the area around the balances clean and put the tops back on any reagent bottle that you may have used.

A copy of these rules is also found on page xviii of this manual. You will be asked to sign that copy of the safety rules and return it along with your check–in sheet to your instructor at the first laboratory period of the term.

Measurements and Significant Figures

Science involves numbers, and chemistry is no exception. In the experiments that you will be doing in this laboratory, you will spend much time recording and working with numbers to obtain a desired result. Manipulation of numbers or "number crunching" uses numbers that you obtain from *measurements, definitions,* and *counting.* You will work with numbers of all three types. Thus you need to know how to record measurements properly and how to "crunch" numbers correctly.

In this section, we will consider first how to record numbers obtained by measurement and what those numbers imply. Then we will look at some simple rules for manipulating numbers in calculations so that the final results will be expressed properly.

Reading the Measuring Device

In an experiment you measure something called quantities; that is, you assign a numerical value to something that has magnitude. The quantities that you will measure and the instruments and equipment you will use to take the measurements in this laboratory are listed below:

Quantities & Measuring Devices

Quantity	Measuring Device
mass	electronic balance
volume	buret graduated cylinder
temperature	thermometer
light absorbance	spectrophotometer

Reading the measuring device correctly and recording the number properly is essential in any scientific or engineering laboratory. *The number you enter into your laboratory notebook says something about the instrument or equipment used to take the measurement in addition to the actual value of the quantity.* There are several points that need to be kept in mind.

First, a measurement includes a *number* and *units*. A beaker has a mass of 39.31 g, not 39.31. A buret delivers 17.56 mL, not 17.56. The table on the next page lists the common units and their abbreviations that you will use.

Quantities & Units

Quantity	Unit	Abbreviation
mass	gram	g
volume	liter milliliter	L mL
temperature	degree celcius degree kelvin	°C K
light absorbance	(dimensionless)	– – –

Second, the number part of a measurement includes all of the certain digits *and* the first uncertain digit. This concept needs a little more explanation.

Consider the volume of liquid in a buret and graduated cylinder shown below. The volume is measured from the lowest part of the curved surface, which is called the meniscus. In the buret, the meniscus is between 21.2 mL and 21.3 mL. The certain digits are 21.2. You can (and must!) estimate the distance between the markings on the buret for the last or uncertain digit. In this case, it looks to be 0.8 between the marks. The marks on the buret represent 0.1 mL, and thus the complete measurement is 21.28 mL. For the graduated cylinder, the marks are for 1 mL. Thus the certain digits are 81, and the uncertain one is 0.6. The complete measurement is 81.6 mL.

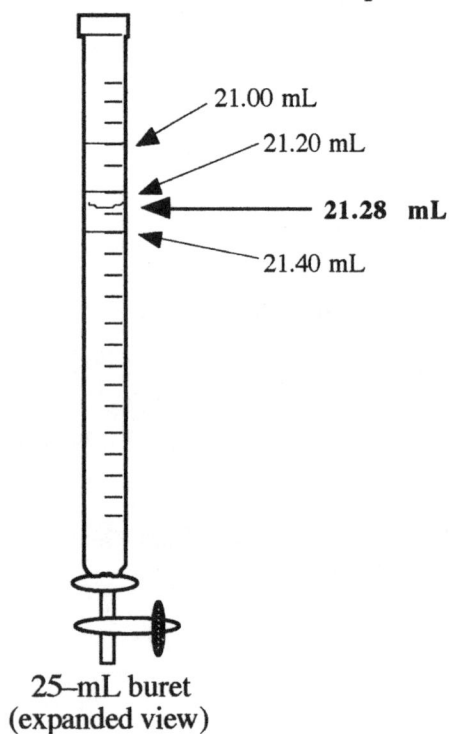

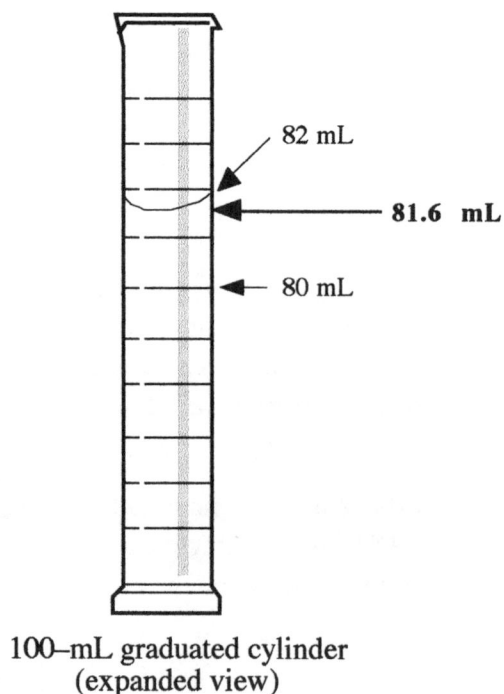

21.00 mL
21.20 mL
21.28 mL
21.40 mL

82 mL
81.6 mL
80 mL

25–mL buret
(expanded view)

100–mL graduated cylinder
(expanded view)

If the meniscus falls exactly on a line, zeros must be included. If the volume in the buret fell exactly on the 22–mL mark, the correct reading would be 22.00 mL. Likewise, if the meniscus in the graduated cylinder were on the 82–mL line, the correct reading would be 82.0 mL.

The *uncertain* digit simply reflects the accuracy of the measuring device being used. It is a valid and essential part of any measurement, and all measurements are uncertain by at least ± 1 unit in the last place recorded.

Significant Figures

Recall that there are three types of numbers that you will be handling in chemistry—numbers from *measurements, definitions* and *counting*.

When a number has come from a definition or from counting, that number is called an *exact* number. For example, by definition there are exactly 100 cm in one meter, or by counting there are exactly 2 atoms of hydrogen and 1 atom of oxygen in 1 molecule of H_2O.

Numbers that come from measurements, however, are not exact and have some degree of uncertainty associated with them. There is no simple name attached to these numbers, but you could call them *uncertain* numbers. As discussed above, the measuring device (such as a balance, buret, or thermometer) reflects this uncertainty, and there is at least one unit of uncertainty in the last digit.

This last digit is *estimated* when reading burets or any device with a scale. For digital–readout instruments, the last recorded digit is *assumed* uncertain. For example, if a crucible gives a mass of 21.857 g on a digital–readout balance, the actual mass is between 21.856 g and 21.858 g. The last "7" is uncertain to ±1. Some laboratory balances are accurate to the nearest 0.01 g, others to 0.001 g, and still others to 0.0001 g. The uncertainty in a measurement of mass would depend on which balance was used, and what you record on your laboratory report tells something about the instrument used to take the measurement.

Because of this uncertainty in measurements, a method has been devised to estimate the "correctness" of calculations that involve measurements. This system uses the term *significant figures* or *significant digits* to refer to the importance or validity of the reported number.

Some rules for identifying the number of significant figures in a number obtained by measurement are given below, and also listed are some rules for determining the number of significant figures after a calculation involving addition, subtraction, multiplication and division. These rules will help you know how much confidence you can place in a calculated value.

Remember that significant figures refer only to *numbers obtained by measurements* and to *calculations involving measurements*! Exact numbers are considered to have an infinite number of significant figures when they are used in calculations.

Rules for Determining Significant Figures

	Examples	No. Sig. Figures
◊ *All nonzero digits are significant.*	3.256 g	4
	428 g	3
◊ *Zeros between nonzero digits are significant.*	105°C	3
	0.2056 g	4
	3.008 mol	4
	70.024 kg	5
◊ *Zeros to the left of the first nonzero digit are not significant.*	0.25 km	2
	0.0035 g	2
	0.00004 atm	1
◊ *Zeros are significant when they appear at the end of a number that includes a decimal point.*	2.570 kJ	4
	82.0 mL	3
	0.6830 g	4
	30. mL	2
◊ *Zeros at the end of a number that does **not** include a decimal point are ambiguous. Expressing a number in scientific notation eliminates the ambiguity.*	250 cm	2 or 3
	2.5×10^2 cm	2
	2.50×10^2 cm	3
	500 g	1, 2 or 3
	5×10^2 g	1
	5.0×10^2 g	2
	5.00×10^2 g	3

Rules for Mathematical Operations

◊ Addition & Subtraction

When numbers involving significant figures are added or subtracted, the result is expressed by rounding the number to the place (before or after the decimal point) with the greatest uncertainty.

You can apply this rule most easily by aligning the numbers in columns (at least in your mind if not on paper). *Remember: Your calculator cannot keep track of significant figures and can easily mislead you—particularly when adding and subtracting!* Also note that you can *lose* or *gain* significant figures when subtracting or adding.

Examples

(1)   0.2865 − 0.015 g = 0.272 g

When you align these numbers in columns, you obtain 0.2715 g. Round this number to 0.272 g because the uncertainty in the third place of 0.015 g is greater than the fourth place of 0.2865 g.

$$
\begin{array}{r}
0.2865 \text{ g} \\
- \; 0.015 \text{ g} \\
\hline
0.2715 \text{ g}
\end{array}
$$

(2)   82 mL − 4 mL = 78 mL

When you align these numbers in columns, you can see no rounding is necessary because the uncertainty is in the units place of both 82 mL and 4 mL.

$$
\begin{array}{r}
82 \text{ mL} \\
- \; 4 \text{ mL} \\
\hline
78 \text{ mL}
\end{array}
$$

(3)   38.45 g + 1.689 g = 40.14 g

Align these numbers in columns, and you can see that the greater uncertainty is in the second place of 38.45 g.

$$
\begin{array}{r}
38.45 \text{ g} \\
+ \; 1.689 \text{ g} \\
\hline
40.139 \text{ g}
\end{array}
$$

(4)   6.002 kJ − 5.998 kJ = 0.004 kJ

Note how subtraction may cause the loss of significant figures; only one significant figure is in the answer now!

$$
\begin{array}{r}
6.002 \text{ kJ} \\
- 5.998 \text{ kJ} \\
\hline
0.004 \text{ kJ}
\end{array}
$$

(5)   8.3 mL + 2.5 mL = 10.8 mL

Note how addition may increase the number of significant figures. There are three significant figures in the answer but only two in each factor.

$$
\begin{array}{r}
8.3 \text{ mL} \\
+ \; 2.5 \text{ mL} \\
\hline
10.8 \text{ mL}
\end{array}
$$

◊ Multiplication and Division

When numbers involving significant figures are multiplied or divided, the result is expressed by rounding the answer to the number of significant figures as found in the factor with the fewest significant figures.

Examples:

(1) $(12.0 \text{ cm}) \times (0.18 \text{ cm}) = 2.16 \text{ cm}^2 = 2.2 \text{ cm}^2$

The factor 12.0 cm has *three* significant figures but 0.18 cm has only *two* significant figures, and the result is thus rounded to two significant figures.

(2) $(9.862 \text{ g}) \div (10.1 \text{ mL}) = 0.976436 \text{ g/mL} = 0.976 \text{ g/mL}$

You round to three significant figures because the factor with the fewest significant figures (10.1 mL) has only three significant figures.

◊ Combination of Operations

The standard order of mathematical operations is followed when using numbers involving significant figures. Rounding to the correct number of significant figures may be done at each step of the calculation *or* the final result may be rounded to reflect the correct number of significant figures. Remember that your calculator cannot do your thinking for you when it comes to determining the number of significant figures in a calculation.

When adding or subtracting, most students find it helpful to determine (or at least keep in mind) the number of significant figures *after* each step in a multi–step calculation such as the following example.

$$\textit{Example:} \quad \frac{653 \text{ kJ} - 602 \text{ kJ}}{602 \text{ kJ}} \times 100\% = 8.5\%$$

This result is rounded to only two significant figures because the numerator becomes 51 kJ after the subtraction (the first mathematical operation). A key to this type of calculation is to determine the number of significant figures *after the subtraction* because that value must be reflected in your final result.

◊ Rounding

Consider three numbers (or places) *beyond* the place to which you are rounding. When these numbers (real or implied) to be dropped are greater than 500, add 1. When the numbers to be dropped are less than 500, leave the number alone. When the numbers to be dropped are exactly 500, round to an even number.

Examples: (the underlined numbers are beyond the place to which you are rounding)

15.8<u>9</u> cm rounded to three significant figures would be 15.9 cm.

0.258<u>497</u> g rounded to three significant figures would be 0.258 g.

3.4<u>500</u> m rounded to two significant figures would be 3.4 m because *4* is even.

0.7<u>500</u> g rounded to one significant figures would be 0.8 g because *7* is odd.

30.9<u>500</u> g rounded to three significant figures would be 31.0 g because *9* is odd.

Evaluating Your Data

In some experiments that you will do in this laboratory, you will measure one or more quantities (such as mass and volume in Experiment 1) and then use these values to calculate a result. If you have taken and recorded the measurements properly, used the correct equations to calculate a result, and followed the rules for use of significant figures, then how can you evaluate your answer? How "good" is the number you have obtained when you consider that all measurements have some degree of uncertainty associated with them?

As discussed in the preceding section entitled *Measurements and Significant Figures*, every physical measurement involves some degree of uncertainty. This uncertainty arises from the nature of the devices used to make measurements. It can be minimized but never completely eliminated. In all cases, you can, however, evaluate your results.

Two terms apply to values determined by experiment—accuracy and precision. *Accuracy* refers to how close your result is to an accepted or so–called true value. *Precision* reflects how closely repetitive experiments lead to the same value. Your goal in doing any laboratory is to be both precise and accurate.

Precision can be evaluated comparing the deviation of each trial to the average (called the mean) of several trials. Another measure of precision is to calculate the standard deviation of a set of trials. Because you do not have sufficient time to do repetitive experiments in the laboratory periods, you will not have to evaluate the precision of your work. The topics of calculating *relative deviation from the mean* and *standard deviation*, however, are worthwhile, and you may have used them in other courses. Thorough discussions are usually found in books on statistics and analytical chemistry. Also many hand–held calculators can calculate the standard deviation of a set of values.

Accuracy is frequently evaluated by determining what is called relative error or more frequently *percent error*. The formula for this calculation is

$$\% \text{ error } = \frac{(\text{your value} - \text{accepted value})}{(\text{accepted value})} \text{ x } 100\%$$

Accepted values are usually found in the laboratory manual or can be looked up in chemistry handbooks. The percent error may be a positive or negative number indicating whether your result represents a value above or below the accepted value. *Students often make errors in significant figures in calculating the percent error.* Review the calculation shown in the example for *Combination of Operations* on page xii to avoid these common mistakes.

Equipment and Supplies

You will be assigned a drawer called a *station* at the first laboratory of the term. The stations are numbered, and you will work at this station during the term. Your drawer contains equipment and supplies that you will use routinely in the laboratory, and you are responsible for seeing that all of the required equipment is kept in the drawer during the term. Your drawer will be unlocked at the begining of the period and locked as you leave.

A few items are too large to fit in individual desk drawers, and this equipment is found in a drawer at each station labeled *Common Equipment*. Also, some special equipment for a specific experiment may be set out in the common areas. The common and special equipment is *not* to be stored in your laboratory drawer but to be returned to where you obtained it at the end of the day.

Below is a list of what should be kept in each drawer. A check–in sheet and set of safety rules follow. During the first laboratory period, you will be asked to check the items in your station drawer, sign the saftey rules, and turn in these sheets to your instructor.

Laboratory Drawer

Beakers (1 each)
    50–mL, 100–mL,
    150–mL, 250–mL,
    400–mL

Funnel

Evaporating dish

Test tubes (10)

Mortar and pestle

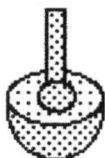

Flasks (3)
    125–mL

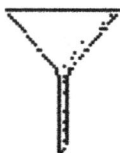

Graduated cylinders
    10–mL, 100–mL

Glass stirring rod

Test tube rack, brush, and holder

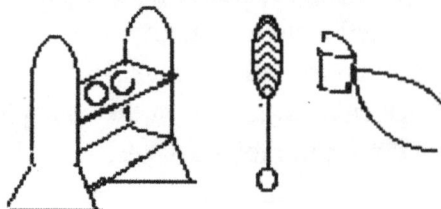

Spatula

Crucible tongs

Clay triangle

Wire gauze

Plastic weigh boats (2)

Medicine dropper

Bunsen burner

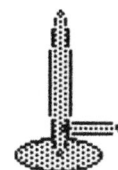

Sparker or matches

Litmus paper (red and blue or neutral)

Common Equipment

Clamp

Clamp holder

Iron ring

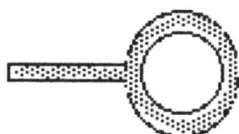

Ring stand (usually found
under the sinks)

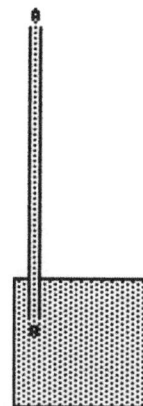

Equipment Check–List

Name: _____

Lab Day & Time: _____   Lab Station Number: _____

Your laboratory desk drawer should contain the items listed below.

	Check–in		Check–out
	Have	*Need*	
50–mL beaker	____	____	____
100–mL beaker	____	____	____
150–mL beaker	____	____	____
250–mL beaker	____	____	____
400–mL beaker	____	____	____
125–mL flask (3)	____	____	____
10–mL graduated cylinder	____	____	____
100–mL graduated cylinder	____	____	____
funnel	____	____	____
evaporating dish	____	____	____
glass stirring rod	____	____	____
test tubes (10)	____	____	____
test tube holder	____	____	____
test tube brush	____	____	____
test tube rack	____	____	____
mortar and pestle	____	____	____
medicine dropper	____	____	____
spatula	____	____	____
crucible tongs	____	____	____
clay triangle	____	____	____
wire gauze	____	____	____
plastic weigh boats (2)	____	____	____
red and blue litmus paper	____	____	____
sparker or matches	____	____	____

Lab Instructor's Initials   *Check–in:* _____   *Check–out:* _____

Safety Rules for the Laboratory

1. You must wear approved safety eye wear at all times in the laboratory.

2. Note and record the location of essential items of safety equipment:

   a. fire extinguishers _____

   b. eye wash stations _____

   c. safety shower _____

3. Note the location of exits from the laboratory.

4. No smoking, eating or drinking is allowed in the laboratory.

5. *Common sense* is the basic safety rule of all chemistry laboratory work.

6. Be careful with the glassware, and never use cracked or broken glassware. Discard broken glass in the special container available in the laboratory.

7. When using a flame, make certain that no flammable materials (including long hair) are nearby. When heating a test tube, never point the tube at anyone or look directly into it. The contents can boil rapidly and "bump" out of the tube.

8. Treat all chemicals as if they were toxic and hazardous. Although you will not be working with any extremely dangerous substances, never taste a chemical, wash any chemicals off your hands with soap and water, and immediately clean up any spills on the counter tops. Avoid rubbing your eyes while working in the laboratory. If you are instructed to smell a chemical, fan the vapors toward your nose and sniff gently. Be careful to use only the specified amounts of any chemical. Also get in the habit of washing your hands shortly after leaving the laboratory for the day.

9. Discard excess chemicals and waste products in the proper containers as directed by your laboratory instructor. Never pour solutions down the sink without permission.

10. Because cleanliness and safety are related, keep your work area in order. Also keep the area around the balances clean and put the tops back on any reagent bottle that you may have used.

I have read the above safety rules and agree to wear the proper safety eye wear and to abide by these rules.

Name (printed) _____

Signature _____ *Date* _____

Physical Properties

Purpose

The primary objective of this experiment is to measure the density of solids and liquids. The secondary objectives, which are extremely important in scientific and technical laboratories, are to learn how to take measurements properly and to express calculations with the correct number of significant figures.

Equipment Needed

10–mL and100–mL graduated cylinders, 50–mL beaker, 100–mL beaker, dropper

Reagents Needed

deionized water, 95% ethanol (denatured), samples of metals

Discussion

Chemistry is the science of matter, or in other words, it is the study of substances that make up the universe. One aspect of chemistry is to describe characteristics of the substances that are found around us. Chemists call the attributes of substances *properties,* and they divide properties into two general types—*physical* and *chemical.*

Physical properties are ones that you can measure without producing a new substance in the process of taking the measurement. For example, temperature is a physical property. You can measure the temperature of a liquid by immersing a thermometer in it. This procedure does not change the liquid into anything else, and hence, temperature is a physical property. An expanded list of physical properties includes color, density, melting point, boiling point, crystalline form, hardness, malleability, ductility, thermal conductivity, electrical conductivity, solubility, odor, and specific heat.

Chemical properties are ones that do involve making a new substance in the process of taking the measurement. For example, you determine the reactivity of hydrogen with oxygen by carrying out a reaction in which water is formed. Another example is the observation that iron exposed to air forms rust on its surface. Rust is a compound of iron and oxygen, and thus a new substance is involved in determining this property of iron. You will study chemical properties of various substances in a later experiment.

The physical property that you will determine in this experiment is density. Density is the ratio of mass to volume and is a useful number describing pure substances and homogeneous mixtures.

$$\text{density} = \frac{\text{mass}}{\text{volume}}$$

The International System (called SI) unit for density is kg/m^3, but chemists find a more useful set of units are g/cm^3 or g/mL (since 1 mL = 1 cm^3). The g/mL term is frequently used for liquids, and the g/cm^3 is reported for solids.

Densities of pure metals fall into two general classes: light metals with densities of < 3 g/cm^3 and heavy metals with densities > 3 g/cm^3. Light metals are in the representative groups of the periodic table (groups 1A, 2A) and Al, whereas heavy metals are the transition metals (the B groups) and the post–transition metals Ga, In, Sn, Tl, Pb and Bi. The lightest metal is Li, 0.534 g/cm^3, and the heaviest are Os and Ir, 22.6 g/cm^3. Heavy metals have densities about 7 g/cm^3 and above with the exception of titanium which has a density of 4.51 g/cm^3. A table at the end of this experiment lists densities of some common metals.

Pure liquids have densities close to 1 g/mL. Typical values of some common liquids range from diethyl ether, 0.708 g/mL, to carbon tetrachloride, 1.604 g/mL. A notable exception to this range of densities is Hg, which is the only metal that is a liquid at ordinary temperatures. The density of Hg is 13.6 g/mL.

Density varies with temperature, and precise calculations require that you state the temperature at which the measurements are taken. Temperature effects on density are more pronounced with liquids than with solids, although expansion and contraction of metals with temperature changes are of concern to engineers in the construction of bridges and buildings. In this experiment, you will not consider the effects of temperature on density determinations.

Procedure

I. Density of a solid by approximate method

A. *Measurements and data collection*

The top–loading balances measure to the nearest 0.001 g up to 25–30 g and then to the nearest 0.01 g above 30 g. Make certain that you zero the balance before each measurement by pushing the tare bar until three zeros appear to the right of the decimal point. *It is essential that you measure each mass to the maximum number of digits that the balance will allow!* Also, always use the same balance when weighing a particular object more than once.

Weigh a sample of metal on the laboratory balances and record your measurement on the report sheet. Use about 30 g of a light metal or 120 g of a heavy metal, and note the metal that you use on your laboratory report.

Add about 40–50 mL of deionized water to the 100–mL graduated cylinder. Measure the volume to the nearest 0.1 mL. The last digit in this measurement is estimated as discussed in the section on *Measurements and Significant Figures* of this manual.

Carefully add the metal sample to the graduated cylinder without splashing out any water. Record the new volume to the nearest 0.1 mL. If the water does not completely cover the metal sample, empty the cylinder and repeat the volume measurements after adding more water to the cylinder than in your first trial.

B. *Calculations*

Calculate the volume of the metal by subtracting the two volumes measured in the graduated cylinder. This difference is the volume of the metal.

Calculate the density of the metal by dividing the mass of the metal sample by its volume. This number is the experimental value (referred to below as "your value").

Determine the percent error of your calculation using the general formula

$$\% \text{ error} = \frac{(\text{your value} - \text{accepted value})}{(\text{accepted value})} \times 100\%$$

A table on page 4 lists accepted values for common metals and liquids. The percent error may be either positive or negative.

II. Density of a liquid

A. *Measurements and data collection*

Add about 15 mL of 95% ethanol from the storage bottle into a 100–mL beaker. Carefully add 10 mL of this ethanol to a 10–mL graduated cylinder. Use the dropper to get the meniscus exactly on the 10–mL mark. This volume is exactly 10.00 mL.

Weigh a clean, dry 50–mL beaker on a top–loading balance. Transfer the measured 10.00–mL volume of ethanol to the weighed beaker, and measure its combined mass on the balance.

B. *Calculations*

Calculate the mass of the liquid by subtracting the two masses of the 50–mL beaker, and determine the density by dividing this mass by the volume of ethanol.

Obtain the accepted value of the density of ethanol from the table below, and calculate the percent error as before.

III. Cleanup

Dry and return the metals to the appropriate bottles. Discard any unused ethanol in the waste bottle.

IV. Review of metals and working with significant figures

Answer the questions concerning metals and complete the significant figure exercises on the laboratory report. You may have to use your textbook in your review of metals.

Density of Common Substances at 25°C

Light *less* 3g cc
Heavy *more*

Substance	Formula	Density g/cm³ or g/mL
Metals		
aluminum	Al	2.70
cadmium	Cd	8.64
chromium	Cr	7.20
cobalt	Co	8.9
copper	Cu	8.92
gold	Au	19.3
iron	Fe	7.86
lead	Pb	11.3
magnesium	Mg	1.74
manganese	Mn	7.47
mercury	Hg	13.6
nickel	Ni	8.90
palladium	Pd	21.4
platinum	Pt	21.4
silver	Ag	10.5
tin	Sn	7.28
titanium	Ti	4.51
tungsten	W	19.4
zinc	Zn	7.14
Liquids		
acetic acid	$HC_2H_3O_2$	1.046
acetone	CH_3COCH_3	0.791
carbon tetrachloride	CCl_4	1.604
ethanol (95%)	C_2H_5OH	0.800
octane	C_8H_{18}	0.703
toluene	C_7H_8	0.867

Separation of a Mixture

Purpose

The objective of this experiment is to separate a mixture into its individual components. You will use several physical methods and examine the properties of three types of solids.

Equipment Needed

100–mL beaker, 150–mL beaker, 250–mL beaker, evaporating dish, wire gauze, 10–mL graduated cylinder, funnel, Bunsen burner, stirring rod, ring stand assembly, filter paper

Reagents Needed

mixture containing iodine, sodium nitrate and silicon dioxide; ice

Discussion

A mixture contains two or more pure substances (elements and/or compounds) that are not chemically combined. Each component in a mixture retains its own chemical identity, and when isolated from a mixture, each component is pure and unchanged.

Mixtures are divided into two types—heterogeneous and homogeneous. Heterogeneous mixtures are not uniform throughout. Composition varies within the mixture itself as evidenced by concrete in which gravel settles to the bottom before the concrete hardens. Homogeneous mixtures, on the other hand, are uniform throughout. A solution is an example of this type of mixture. In a salt solution, the concentration of ions is the same at any location in the solution.

Mixtures also have variable compositions, but particularly for solutions, there are usually limits to the amounts of substances than can be mixed and still produce a homogeneous mixture. Only so much sugar can dissolve in water forming a homogeneous mixture; any additional sugar will not dissolve and will settle to the bottom of the container.

Substances in nature usually occur in mixtures; very few substances are found in the pure state. Thus in many industrial and laboratory procedures, chemists and engineers face the task of separating mixtures into pure components.

Separation techniques are physical processes in which components are separated by differences in their properties. For example, sodium chloride is soluble in water but calcium carbonate is not. The difference in solubility in water (a physical property) implies that you can separate a mixture of the two salts by adding water. Calcium carbonate would not dissolve and could be removed by filtration. Sodium chloride could be recovered by evaporating the water.

Typical physical processes used for separations are solubility, distillation, filtration, sublimation, crystallization, migration in an electric field, and sedimentation. Techniques based on these processes are called such things as chromatography, fractional distillation, fractional crystallization, electrophoresis and centrifugation. You may learn about these processes and the methods that use them in other courses in chemistry and engineering.

Four types of substances exist in the solid state, and this classification reflects the nature of the structural units and forces (called chemical bonds) that hold the structural units of the solid together. These four types are: (1) ionic solids, (2) molecular solids, (3) covalent or macromolecular solids, and (4) metallic solids. The table below lists some *general* characteristics of these solids.

Characteristics of Solids

Type	Structural Units	Bonding	Properties
ionic	cations/anions	ionic bond	hard, brittle, high mp, poor conductor, soluble in H_2O
molecular	molecules	intermolecular forces	soft, low mp, poor conductor, soluble in organic solvents
covalent	molecules	covalent bonds	very hard, very high mp, insulator, insoluble
metallic	atoms	metallic bond	soft to hard, low to high mp, good con– ductor, insoluble

In the previous experiment, you studied some properties of metals, and in this experiment, you will work with the other types of solids. You will separate a mixture of iodine (a molecular substance), sodium nitrate (an ionic compound), and silicon dioxide (a covalent solid).

A diatomic molecule, I_2, forms the structural unit of solid iodine. The electron cloud surrounding the two iodine nuclei in the molecule deforms and generates temporary regions rich in negative electrons and deficient in negative electrons (and hence a positive region). Electrostatic attractions between these temporary positive and negative regions of the electron clouds on adjacent iodine molecules are called dispersion forces, and these dispersion forces hold iodine molecules together in the solid. Sodium nitrate, $NaNO_3$, contains the ions Na^+ and NO_3^- in an equal ratio, and the electrostatic attraction of the ions maintains the compound in the solid state. Silicon dioxide has the empirical formula SiO_2 but is a giant network in which each silicon atom is bonded to four oxygen atoms, and each

oxygen atom is bonded in turn to two silicon atoms. The name of this type of _ _ _ _ _ _
covalent because the structure is one giant, covalently bonded network of atoms. Sea sand is
mostly SiO_2. You will learn more about solids and the forces that hold them together later
in your study of chemistry.

Ionic, molecular and covalent substances can frequently be distinguished by their melting
points and their solubility in water and organic solvents. Molecular compounds have low
melting points, are poorly soluble in water but readily soluble in organic solvents. Ionic
solids have high melting points, and many are soluble in water but do not dissolve in
organic solvents. Covalent compounds have extremely high melting points and are virtually
insoluble in all solvents. You will use some of these characteristic properties in this
experiment.

A flow diagram of the separation procedure for this experiment follows. Iodine sublimes
(goes directly from the solid to the gaseous state) at moderately low temperature at
atmospheric pressure, whereas sodium nitrate and silicon dioxide do not. Thus heating the
mixture removes iodine by sublimation. When iodine vapors contact the cold surface of an
evaporating dish, solid iodine is deposited on its surface and can be recovered in pure form.

Water is added to the remaining mixture. Sodium nitrate dissolves readily, and the insoluble
silicon dioxide is removed from the solution by filtration. Evaporation of the water leads to
the crystallization of sodium nitrate. By the above physical processes, the mixture is
separated into its individual components.

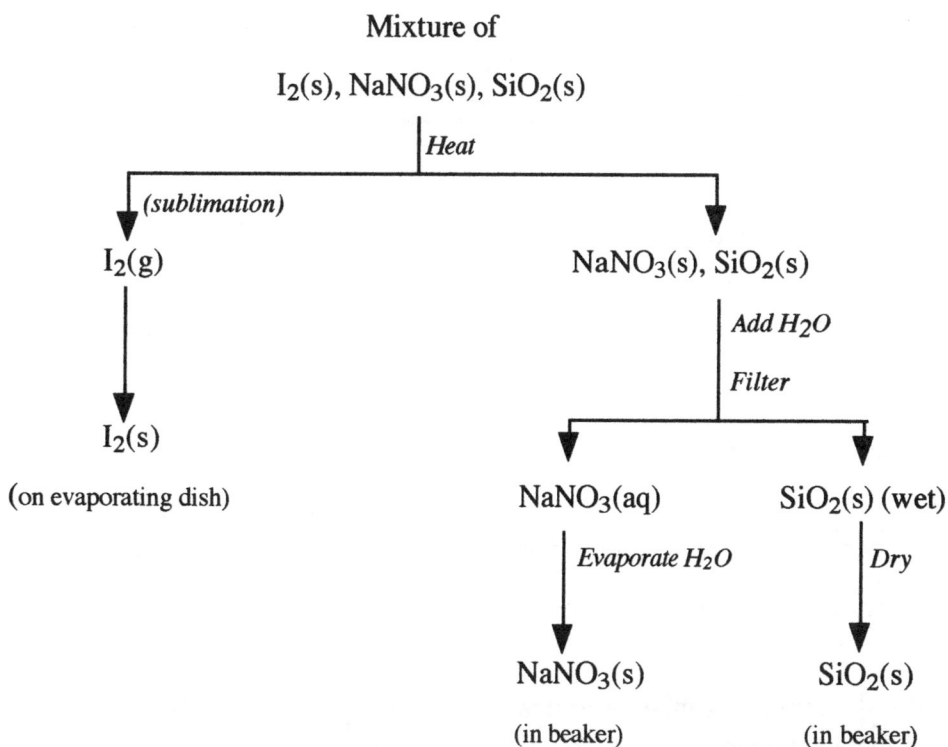

Mixture of

$I_2(s)$, $NaNO_3(s)$, $SiO_2(s)$

| *Heat* |

(sublimation)

$I_2(g)$ $NaNO_3(s)$, $SiO_2(s)$

| *Add H_2O* |

| *Filter* |

$I_2(s)$

(on evaporating dish) $NaNO_3(aq)$ $SiO_2(s)$ (wet)

| *Evaporate H_2O* | | *Dry* |

$NaNO_3(s)$ $SiO_2(s)$

(in beaker)   (in beaker)

Procedure

I.   Separation of iodine

Weigh a clean, dry 100–mL beaker to the nearest 0.01 g. Add about 5 g of the mixture to this beaker. Reweigh the beaker and its contents to the nearest 0.01 g.

Place the beaker on wire gauze on a ring stand as shown below. Cover the beaker with an evaporating dish that contains several pieces of ice. Heat the bottom of the beaker gently with a small flame. Observe the changes taking place. Continue heating until the cloud of iodine vapor has disappeared and iodine is deposited on the surface of the evaporating dish. Keep ice in the evaporating dish during this procedure.

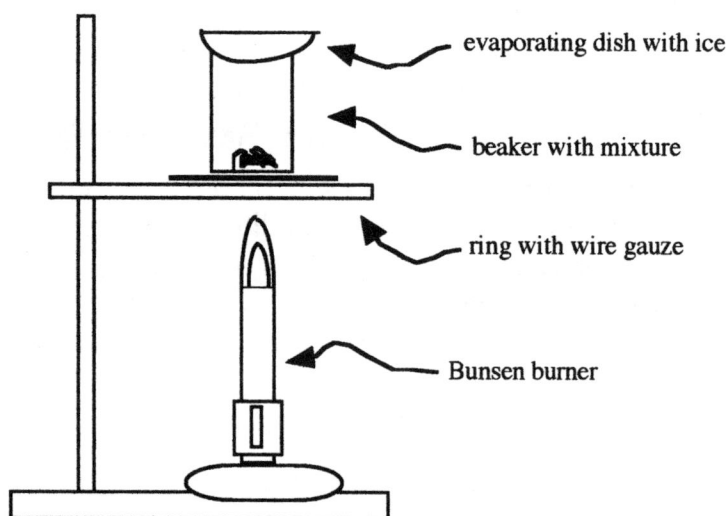

- evaporating dish with ice
- beaker with mixture
- ring with wire gauze
- Bunsen burner

Allow the beaker to cool, and remove the evaporating dish carefully. Pour the water and ice into the sink. Scrape the deposited iodine into a beaker provided in the hood.

Reweigh the cool beaker. It now contains only two components of the mixture.

II.   Separation of sodium nitrate

Pour about 10 mL of water into the beaker containing the solid residue. Stir for several minutes with a glass stirring rod until the sand swirls easily around the beaker.

Weigh a 150–mL beaker to the nearest 0.01 g. Set up a funnel with filter paper folded as shown in the figure on the next page with the 150–mL beaker under the funnel.

Pour the remaining mixture into the funnel using the stirring rod as a guide. Rinse out into the funnel any remaining silicon dioxide with a few more milliliters of water.

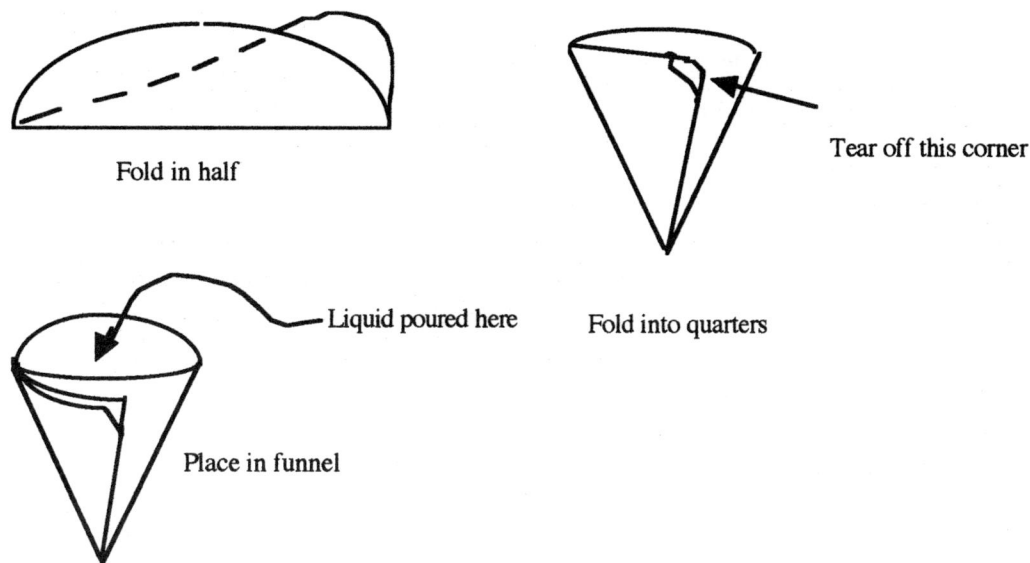

Fold in half

Tear off this corner

Liquid poured here

Fold into quarters

Place in funnel

Place the 150–mL beaker on wire gauze on a ring stand, and boil away the water with a low flame. When the water is almost gone, be careful not to splatter the remaining solution out of the beaker. You may have to remove the flame temporarily if the solution begins to splatter extensively.

Cool the beaker, and reweigh as before.

III.   Isolation of silicon dioxide

Weigh a clean, dry 250–mL beaker to the nearest 0.01 g. Gently scrape the sand (silicon dioxide) off the filter paper into the beaker, and heat the beaker gently on the ring stand. When the water has been removed by evaporation, cool the beaker and reweigh it. You may also have to use care in evaporating the water without extensive splattering.

IV.   Calculations

Determine the mass of iodine present in the mixture by the difference in the mass of the mixture before and after sublimation of the iodine.

Determine the amount of sodium nitrate present from the mass of solid in the 150–mL beaker after the water has been evaporated.

Determine the mass of silicon dioxide present from the mass of solid in the 250–mL beaker.

Convert the mass of each substance present into a percent by mass of the total mixture. Calculate the recovery of pure substances from the mixture by adding the percent values.

Recheck all your calculations for the correct number of significant figures. You may need to review how many significant figures are recorded when you divide and add numbers involving measurements.

V.   Cleanup

The iodine that you isolate should be scraped into the covered beaker in the hood. Discard the sand and filter paper into the waste containers, and rinse the sodium nitrate into the labeled waste beaker. This solution is a mild nitrogen fertilizer. Wash all glassware. Return the common equipment to the proper location.

VI.   Review of pure substances

Answer the set of questions on the second page of the laboratory report concerning pure substances, elements and compounds.

Empirical Formula of a Compound

Purpose

The objectives of this experiment are to synthesize a compound and determine its empirical formula. A secondary purpose is to represent chemical changes with words and symbols.

Equipment Needed

crucible and cover, crucible tongs, ring stand with ring, clay triangle, burner, sparker

Regents Needed

magnesium ribbon

Discussion

Chemical compounds are generally divided into two major classes—molecular and ionic. A molecule is the smallest structural unit of a molecular compound, and it consists of an aggregate of atoms held together by forces called chemical bonds. Ionic compounds contain cations and anions that are attracted to each other by electrostatic forces. The simplest ratio of these ions that composes an electrically neutral unit is sometimes called a formula unit, and these formula units represent the smallest structural entity of ionic compounds.

A chemical formula indicates the elements present in a compound (whether molecular or ionic) and gives you information about the ratio of atoms or ions present in that compound. Thus a formula symbolically represents a compound and tells something about its composition. You will encounter three types of formulas in chemistry—empirical, molecular and structural.

An empirical formula represents the simplest ratio of atoms or ions present in the smallest structural unit of a compound. Another common name for empirical formula is the *simplest formula*. As noted above, no discrete unit exists in ionic compounds. The number of cations and anions present in an electrically neutral formula unit is the simplest ratio of ions in the compound, and hence all ionic compounds are represented only by empirical formulas. When empirical formulas are used for molecular compounds, the formula tells you the simplest ratio of each type atom present in a molecule.

A molecular formula gives the actual number of atoms of each element present in a molecule. As the name implies, this formula is used only for molecular compounds. The exact numbers of each type atom may *also give* the simplest ratio of atoms present in a molecule, and in such cases, the empirical and molecular formulas are the same. In other examples, the molecular formula is an integral multiple of the empirical formula.

Structural formulas indicate the exact number of atoms present in a molecule or polyatomic ion and also tell you something about the three–dimensional structure of the unit. The geometry of molecules and polyatomic ions is covered later in your study of chemistry.

In empirical and molecular formulas, the number of ions or atoms in the structural unit is indicated by subscripts following the symbol for the element. In ionic compounds, the cation component is written first followed by the anion. The charge present on the ions *is not written* in complete formulas for ionic compounds. You are expected to know the charges from the nature of the elements from which monoatomic ions are derived or from learning the charges on polyatomic ions. For molecular compounds, positions of the elements on the periodic table determine the order in which they are usually written in the formula. The element written first is usually to the left and/or above the element written second.

The table below indicates empirical and molecular formulas for some compounds. Note that only empirical formulas represent ionic compounds and that one of the ionic compounds consists of polyatomic cations and anions. Also note that for some molecular compounds the two formulas are the same.

Comparison of Formulas

Compound Name	Empirical Formula	Molecular Formula
water	H_2O	H_2O
calcium iodide	CaI_2	——
methane	CH_4	CH_4
butane	C_2H_5	C_4H_{10}
ammonium sulfate	$(NH_4)_2SO_4$	——
glucose	CH_2O	$C_6H_{12}O_6$

In this experiment, you will synthesize a compound containing magnesium and oxygen and determine its empirical formula from mass measurements. The reaction between magnesium metal and oxygen in the air does not occur rapidly at room temperature but does so at the temperature of a gas burner (about 700°C). A small ceramic bowl called a crucible withstands such temperatures, and simple reactions of this type are carried out in crucibles over a flame.

The magnesium metal to be used is in the form of a ribbon so that there is a large surface area on which the reaction may occur. Oxygen for the reaction is simply that in the air.

Another reaction also occurs to a limited extent when air is mixed with ma
elevated temperatures. This reaction is between magnesium and nitrogen fr
Fortunately, a simple reaction exists to convert the undesirable magnesium–nitrogen
compound back into the magnesium–oxygen compound. This reaction is carried out in the
crucible and takes place when water is mixed with the magnesium–nitrogen compound at
high temperature. Products of the reaction are the desired magnesium–oxygen compound
and ammonia, which is volatilized at the high temperature of the conversion.

A secondary objective of this experiment is to give you practice in naming compounds and
expressing chemical reactions both in written statements and with chemical formulas. Thus
you will need to know the names and formulas of magnesium metal, oxygen and nitrogen
as found in the air, water, ammonia, the magnesium–oxygen compound, and the
magnesium–nitrogen compound.

Procedure

I.   Formation of magnesium–oxygen compound

Rinse and clean the crucible and its cover with water. Set up the ring stand with ring and
clay triangle. Place the crucible and cover on the clay triangle.

Heat the empty crucible with the cover tilted for 10 minutes. The bottom of the crucible
should glow a dull red color. The inner blue cone of the flame should be about 2 cm from
the bottom of the crucible. Cool the crucible and cover to room temperature (about 5
minutes), and weigh to the nearest 0.001 g.

Add magnesium ribbon that has been rolled into a ball about the size of a marble, and weigh
the crucible and cover with the magnesium again.

Heat the covered crucible slowly at first, then at full heat for 15 minutes. Cool for about 1
minute on the ring stand, and then tilt the cover on the crucible and heat for 10 minutes.
Remove the cover and look at the material in the crucible. If it is burning like cigarette paper,
replace the cover at a tilt and heat again for 5 minutes. If the material is not burning but just
glowing, remove the cover and heat for 5 minutes.

Cool the crucible for 5 minutes. Add 10 drops of distilled water to the cooled crucible.
Crush the product with a glass stirring rod and rinse any product off the glass rod with
several more drops of water. Replace the cover at a tilt, heat the crucible slowly at first until
you can see no vapors coming from the crucible, and then heat strongly for 10 minutes.
[Why do you perform this reaction?]

Cool the crucible and cover to room temperature, and weigh the crucible and cover with its
contents.

II.   Calculations

From the three mass measurements, determine the mass of magnesium and oxygen in the magnesium–oxygen compound. These numbers will probably have 3 significant figures.

Calculate the percent by mass of magnesium and oxygen in the magnesium–oxygen compound.

Convert the percent values to moles of each element present in 100 g of compound, and finally determine the mole ratio of magnesium to oxygen that you determined experimentally.

III.   Expressing chemical reactions with formulas and words

Answer the two questions on the back of the laboratory report. Chemists—and students of chemistry—need to express chemical processes in written statements and also symbolically in chemical equations. The first question requires that you write and balance the chemical reactions that occurred within the crucible using the correct chemical formulas. The second question asks you to write grammatically correct sentences that describe these chemical changes.

IV.   Cleanup

Gently scrape the magnesium–oxygen compound that you synthesized into the waste containers. Rinse the crucibles with tap water, and return them to the supply desk.

Chemical Reactions

Purpose

The purpose of this experiment is to perform some chemical reactions. During this laboratory, you will observe some chemical properties. You will also write and balance chemical equations to represent the changes that have occurred and identify the general type of the reaction.

Equipment Needed

10 test tubes, test–tube holder, test–tube rack, crucible tongs, litmus paper (red and blue *or* neutral), 250–mL and 400–mL beakers, burner and sparker, delivery tube, ring stand, clamp and clamp holder

Reagents Needed

potassium chlorate, manganese(IV) oxide, magnesium ribbon, dilute hydrochloric acid, copper(II) carbonate, lime water, aluminum rod or foil, copper(II) chloride solution, calcium metal, calcium chloride solution, sodium carbonate solution, dilute acetic acid, sodium hydrogen carbonate

Discussion

A chemical reaction is a process in which one set of substances is changed into a new set of substances. The substances at the start are called *reactants*, and the substances at the end of a reaction are called *products*. A chemical reaction is also called a chemical change.

Chemical equations are symbolic representations of chemical reactions and are written as

$$Reactants \rightarrow Products$$

The arrow implies that reactants have been converted into products. Writing a chemical equation requires: (1) chemical formulas of all reactants and products and (2) balancing the number of each type of atom on both sides of the equation. These concepts are discussed fully in you textbook, and hopefully you have already acquired the skills of writing and balancing chemical equations.

Chemical changes are detected by observing the disappearance of reactants and/or identifying the products that form. Both physical and chemical properties of reactants and products are used to detect chemical reactions.

In this experiment, you will perform several chemical reactions and identify some of the products by observing a physical or chemical property of the substances formed. Also you will write and balance chemical equations to represent the reactions that you have done. The specific property that you will use to identify products is given in the *Procedure* section for each reaction.

Another aspect of the laboratory is to observe various types of chemical reactions. Chemical reactions are often classified by the nature of the reactants and products *and/or* on a description of the chemical change that occurs. The major types that you will encounter in general chemistry are precipitation reactions, acid–base reactions (also called neutralizations), oxidation–reduction reactions, and simple decomposition reactions. Textbooks in general chemistry usually discuss subtypes of each of these major classifications.

Precipitation reactions are identified by the formation of an insoluble product when two aqueous solutions of ionic compounds are mixed. For example, when you mix aqueous solutions of ammonium sulfide and cadmium nitrate, a yellow solid identified as cadmium sulfide comes out of solution.

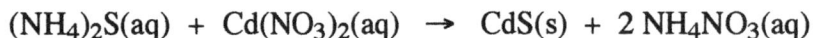

$$(NH_4)_2S(aq) \ + \ Cd(NO_3)_2(aq) \ \rightarrow \ CdS(s) \ + \ 2\,NH_4NO_3(aq)$$

The solid that forms is called a *precipitate*. You can identify these reactions by noting that both reactants are in the aqueous state whereas one of the products becomes a solid. Rules of solubility for ionic compounds can help you predict whether a precipitate will form when various aqueous solutions are mixed.

Acid–base reactions in aqueous systems involve the reaction between an acid and a base to form a salt and often water. To identify these reactions, you must be able to recognize acids, bases and salts. Briefly, an acid is defined as a substance capable of donating a hydrogen ion to another substance, and a base is a substance capable of accepting a hydrogen ion from another substance. Salts are ionic compounds in which the cation is something other than the hydrogen ion and the anion is something other than the hydroxide or oxide ion. Two acid–base reactions are given below. Can you name all the substances involved?

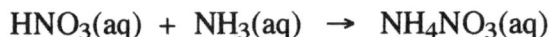

$$H_2SO_4(aq) \ + \ 2\,KOH(aq) \ \rightarrow \ K_2SO_4(aq) \ + 2\,H_2O(\ell)$$

$$HNO_3(aq) \ + \ NH_3(aq) \ \rightarrow \ NH_4NO_3(aq)$$

Sometimes, you may not recognize acids and bases as readily as you can identify the above compounds containing hydrogen and hydroxide ions. For example, nonmetal oxides when mixed with water form acids. Thus when nonmetal oxides are bubbled through any aqueous solution, an acid forms. If a base is also present, then the newly formed acid immediately reacts with it, and overall the reaction produces a salt and water. This two–stage reaction occurs so quickly that usually only the overall reaction is considered. An example involving sulfur trioxide bubbling through an aqueous solution of sodium hydroxide is shown below. You will also encounter this type of reaction in this laboratory.

$$SO_3(g) + H_2O(\ell) \rightarrow H_2SO_4(aq)$$

$$H_2SO_4(aq) + 2\,NaOH(aq) \rightarrow Na_2SO_4(aq) + 2\,H_2O(\ell)$$

Overall reaction: $SO_3(g) + 2\,NaOH(aq) \rightarrow Na_2SO_4(aq) + H_2O(\ell)$

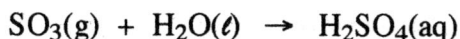

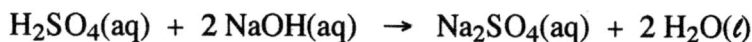

Likewise, a metal oxide forms a base when mixed with water. In the presence of an aqueous acid, an acid–base reaction then occurs and produces a salt and water. An example of this type of overall reaction is shown below for calcium oxide added to an aqueous solution of hydrochloric acid.

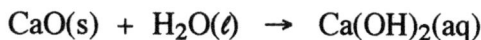

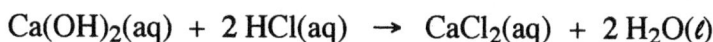

$$CaO(s) + H_2O(\ell) \rightarrow Ca(OH)_2(aq)$$

$$Ca(OH)_2(aq) + 2\,HCl(aq) \rightarrow CaCl_2(aq) + 2\,H_2O(\ell)$$

Overall reaction: $2\,HCl(aq) + CaO(s) \rightarrow CaCl_2(aq) + H_2O(\ell)$

Nonmetal oxides are often called *anhydrous acids*, and metal oxides are referred to as *anhydrous bases*. Just add water, and you get an acid or a base!

Oxidation–reduction reactions (also called *redox reactions*) are another major type of chemical reaction. These reactions are identified by substances that either lose or gain electrons as a reaction proceeds or by elements undergoing a change in oxidation number during a reaction. Oxidation represents a loss of electrons or an increase in oxidation number. Reduction involves a gain of electrons or a decrease in oxidation number. In any redox reaction, one component is always oxidized while another is being reduced. Prior to completing this experiment, you may need to review in your textbook such topics as these basic definitions, how to identify redox reactions and how to determine oxidation numbers.

There are six types of redox reactions that you can identify from the reactants and products involved. Examples of these redox reactions are listed below. Try to note changes in electrons or in oxidation numbers.

1. Combustion reactions and reactions involving *only* elements as reactants

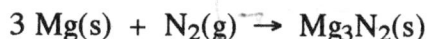

$$CH_4(g) + 2\,O_2(g) \rightarrow CO_2(g) + 2\,H_2O(\ell)$$

$$3\,Mg(s) + N_2(g) \rightarrow Mg_3N_2(s)$$

2. Decomposition reactions producing a free element

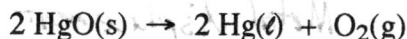

$$2\,HgO(s) \rightarrow 2\,Hg(\ell) + O_2(g)$$

3. Hydrogen displacement from water, steam or acid (A displacement from an acidic solution is shown below.)

$$Zn(s) + 2\,HCl(aq) \rightarrow H_2(g) + ZnCl_2(aq)$$

4. Metal displacement

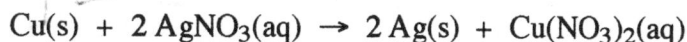

$$Cu(s) + 2\,AgNO_3(aq) \rightarrow 2\,Ag(s) + Cu(NO_3)_2(aq)$$

5. Halogen displacement

$$Cl_2(g) + 2\,KI(aq) \rightarrow I_2(s) + 2\,KCl(aq)$$

6. Disproportionation reaction

$$Cl_2(g) + 2\,NaOH(aq) \rightarrow NaOCl(aq) + NaCl(aq) + H_2O(\ell)$$

Another type of chemical reaction that you often encounter is a **simple decomposition reaction** *without the formation of an element*. A commercially important reaction of this type is the decomposition of calcium carbonate to calcium oxide and carbon dioxide. Calcium oxide is commonly known as lime, and lime has many industrial, commercial and agricultural uses. Note that in this reaction, oxidation numbers do *not* change, and thus this reaction is *not a redox reaction*.

$$CaCO_3(s) \rightarrow CaO(s) + CO_2(g)$$

Procedure

In this experiment, you will be working with flames and generating small amounts of very reactive substances. Please observe the safety rules and take extra care as you do the experiments. Also discard excess chemicals and all products in the proper waste receptacles indicated by your instructor.

Carry out the following reactions and complete the laboratory report as you do the experiments. For each reaction, you will (1) write the chemical formulas of the reactants and products, including the states, (2) write and balance the chemical equation representing the reaction, and (3) classify the reaction as one of the types discussed above. The following list summarizes the reaction types you *may* encounter, and an activity table is given to help you predict and identify products of hydrogen and metal displacement reactions. You may also need to use your textbook or class notes to confirm the chemical formulas for the substances that are involved in these reactions.

Types of Reactions

Precipitation ~ 1
Acid–base — 2
Redox Metal displacement — 1 Combustion R1 Decomposition with formation of an element — 1 Hydrogen displacement — 2 Halogen displacement Disproportionation
Simple decomposition (not a redox) ~ 1

Activity Table

Metal	Displaces H_2 from acid	Displaces H_2 from steam	Displaces H_2 from cold water
Li	X	X	X
K	X	X	X
Ba	X	X	X
Ca	X	X	X
Na	X	X	X
Mg	X	X	
Al	X	X	
Zn	X	X	
Fe	X	X	
Cd	X		
Co	X		
Ni	X		
Sn	X		
Cu			
Hg			
Ag			
Pt			
Au			

REACTION 1 — Place a small quantity (about the size of a pea) of potassium chlorate in a dry test tube. The potassium chlorate contains a small amount of manganese(IV) oxide that acts as a catalyst to accelerate the reaction. Heat the tube strongly until you see a change in the contents and evidence of a gas being generated.

Test the gas for the presence of oxygen by inserting a glowing wood splint into the opening of the test tube. Do not drop the splint into the tube but hold onto it! If oxygen gas is present, the splint will glow brightly or possibly burst into flame. The other product (which you do not identify) is potassium chloride. The manganese compound is not affected by the reaction, and in the balanced equation, it is written over the arrow.

Cleanup: When the tube has cooled, add some tap water and loosen the solid with a test–tube brush. Transfer the solid onto the funnel labeled *Wastes from Reaction 1*.

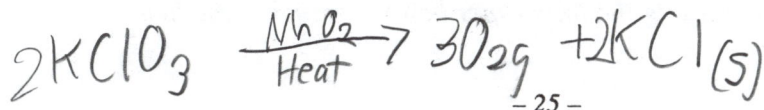

$$2KClO_3 \xrightarrow[\text{Heat}]{MnO_2} 3O_{2g} + 2KCl_{(s)}$$

REACTION 2 — Place a small piece of magnesium ribbon (~5 mm) in about 2 mL of dilute hydrochloric acid in a test tube that is in a test–tube rack. Quickly place another test tube over the mouth of this tube, and collect the gas that is liberated. Try to keep the two tubes tightly sealed together.

When the bubbling has stopped, test the gas in the upper tube for hydrogen as follows. Keep the tube inverted and vertical and rapidly bring a burning wood splint to its mouth. An audible pop or whistling sound indicates the presence of hydrogen gas. The noise is produced by the reaction of hydrogen with oxygen to form water.

In the course of this reaction, magnesium ends up as an aqueous solution of magnesium chloride.

The reaction between hydrogen and oxygen (which may have startled you with its loud pop) releases large amounts of heat per gram of hydrogen, and NASA uses it to power the liftoff of the space shuttle. The unfortunate destruction of the Challenger space shuttle resulted from the uncontrolled reaction of huge amounts of liquid hydrogen and liquid oxygen.

Cleanup: Pour the contents of the test tube into the waste beaker labeled *Wastes for Reaction 2.*

REACTION 3 — Place a small sample of copper(II) carbonate in a test tube that is held at an angle by a ring stand and clamp. Connect a delivery tube as shown below. Add about 10 mL of lime water to the second tube and hold the second tube with a test tube holder so that the end of the delivery tube is beneath the surface of the lime water. Heat the tube with the copper(II) carbonate until bubbles turn the lime water a milky color.

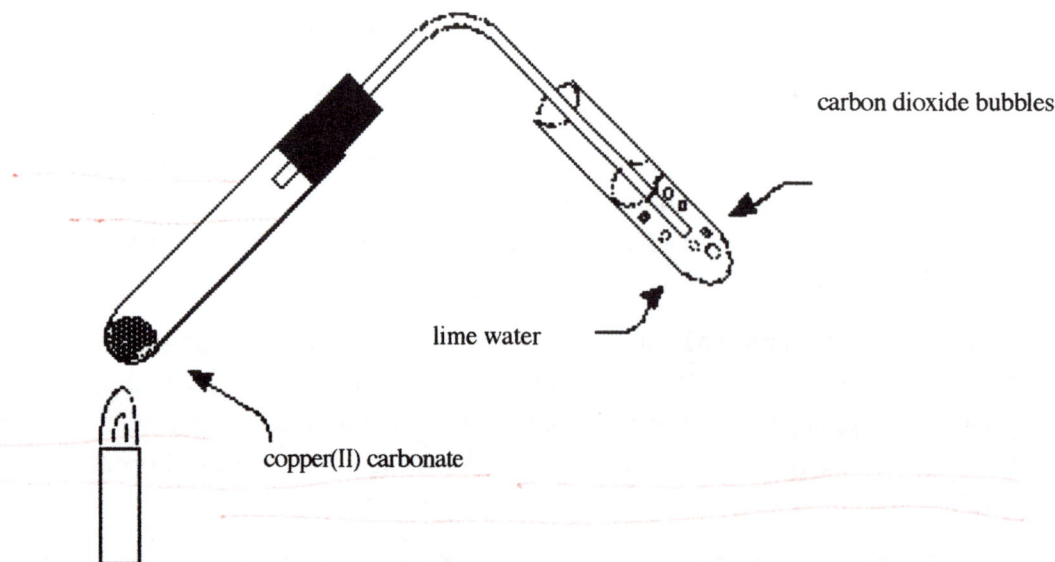

carbon dioxide bubbles

lime water

copper(II) carbonate

Remove the test tube containing the lime water before removing the heat.

Carbon dioxide gas is produced by heating copper(II) carbonate, and carbon dioxide is the substance bubbled through the lime water. The formation of a milky substance in the lime water confirms the presence of carbon dioxide in the bubbles. Lime water is a saturated solution of calcium hydroxide, and the milky substance is a precipitate that forms from a reaction between lime water and a product of carbon dioxide and water. The dark substance that remains in the heated tube is copper(II) oxide.

First you will identify the reaction that took place in the heated test tube, and then you will consider what happened in the lime water. To determine the reaction in the lime water, recall that carbon dioxide is a nonmetal oxide. What type of anhydrous oxide is it—acidic or basic? What happens when one of these oxides is bubbled through water? If that water solution also contains an acid or a base, what other reaction may occur? You may have to review the *Discussion* section of this laboratory to identify the reaction.

Cleanup: After the heated tube has cooled, transfer the *dry* contents to the waste container labeled *Copper(II) Oxide*. Filter the milky lime water into the waste beaker labeled *Lime Water Wastes*.

REACTION 4 — Add about 10 mL of an aqueous solution of copper(II) chloride into a 50–mL beaker. Immerse a piece of aluminum foil or rod into the solution.

Observe the reaction that takes place on the aluminum and any color change that may occur. in the solution. As the reaction proceeds, an aqueous solution of aluminum chloride forms in addition to a metal being plated out of solution.

Cleanup: Transfer all the contents of the beaker to the funnel labeled *Wastes from Reaction 4*.

REACTION 5 — Place a small piece of metallic calcium in about 5 mL of water in a test tube. Set the test tube in a 400–mL beaker. After a few minutes a reaction will occur. Identify one of the products by using the activity table on page 25.

Identify the other product by testing the resulting solution with litmus paper. Red litmus turns blue in basic solutions, and blue litmus turns red in acidic solutions. The compound identified in this manner is only slightly soluble and is the white substance that you see in the test tube.

Cleanup: Transfer the contents of the test tube into the waste beaker labeled *Wastes from Reaction 5*.

REACTION 6 — Obtain about 2 mL of an aqueous solution of calcium chloride in a test tube and about 2 mL of an aqueous solution of sodium carbonate in another test tube. Mix the contents of the two tubes together.

The white solid that forms is calcium carbonate, an insoluble salt. The other ionic compound produced in this reaction is soluble in water.

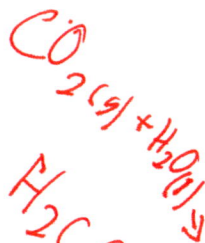

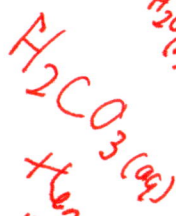

Cleanup: Transfer all the contents of the tube into the beaker labeled *Wastes from Reaction 6.*

REACTION 7 — Add about 2 mL of a dilute solution of acetic acid to a test tube and determine the color that the solution turns litmus paper. Add the contents of the test tube to a small amount of sodium hydrogen carbonate (about the size of a pea) in another test tube. Test the solution with litmus paper after the bubbling has stopped.

The initial products that form are sodium acetate and carbonic acid. Carbonic acid rapidly decomposes into carbon dioxide and water, and the carbon dioxide bubbles out of the solution. In listing the products and writing the balanced equation, consider only the final products after the decomposition of carbonic acid.

To identify the reaction, think about the change indicated by the litmus paper tests. This reaction is similar to the one that you may have used to clean an automobile battery and its terminals.

Cleanup: You may wash the contents of the test tube down the sink with tap water. This reaction has involved two common household products—vinegar (a dilute solution of acetic acid) and baking soda (sodium hydrogen carbonate)!

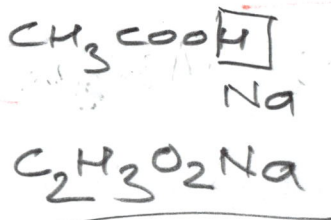

$$CH_3COO\boxed{H}$$
$$Na$$
$$C_2H_3O_2Na$$

Everyday Chemicals

Purpose

The purpose of this experiment is to identify and study some properties of chemicals found in household products. You will examine acid–base and redox properties and perform spot tests for specific ions and compounds.

Equipment Needed

test tubes with rack, spatula, stirring rod, dropper or Beral pipet, 10–mL graduated cylinder, litmus or pH paper, chemistry textbook

Reagents Needed

The following substances will be placed on the reagent shelf: 0.1 M sodium iodide, 0.1 M iodine solution, 1 M sodium hydroxide and 0.1 M hydrochloric acid, universal indicator and starch indicator. Take small quantities of each *as needed.*

Samples of household products will also be available in the laboratory. Obtain small quantities of these substances *as needed.* Do not waste the reagents or the household products.

Discussion

Many commercial products used around the home, garden and workshop contain chemicals that you already know from your introduction to general chemistry. In this laboratory, you will examine some chemical properties of these commercial products and learn how to determine if a particular ion or compound is present in them. The simple experiments and tests that you will do can help you review nomenclature and formulas of ionic and molecular compounds and also help you write and balance chemical equations.

When dissolved in water, many household products produce an **acidic** or **basic** solution. The acidity of the resulting solution then contributes to the action of the substance. For example, washing soda forms a basic solution which helps detergents solubilize grease and dirt. Acidity is easily determined with litmus paper, pH paper, or the colors of the universal indicator. The reason many ionic compounds produce acidic or basic solutions is discussed in your textbook.

Other substances found in household products work through oxidation–reduction reactions such as the oxidizing action of bleach. A simple test for an **oxidizing agent** is to mix the compound with a solution of sodium iodide and look for the appearance of iodine. Oxidizing substances easily remove electrons from I^- ions producing I_2 molecules. Dilute solutions of I_2 appear straw–colored whereas a solution containing sodium iodide is colorless. You can further confirm the presence of I_2 in a solution by adding a few drops of starch indicator and looking for the appearance of the dark blue color that forms when I_2 combines with starch.

Specific ions or compounds can often be identified by simple tests. When compounds containing **ammonia** or the **ammonium** ion are mixed with a strong base such as sodium hydroxide, ammonia is volatilized and can be detected by its odor. Compounds containing the **bicarbonate** or **carbonate** ion are identified by the presence of bubbles of carbon dioxide when the compounds are mixed with an acid such as hydrochloric acid. **Magnesium ions** are detected if a precipitate of magnesium hydroxide occurs when sodium hydroxide is added. As noted above **starch** forms a deep blue color with iodine solutions, and you can use this property to detect the presence of starch. When drops of an iodine solution come in contact with starch, a deep blue color appears. Over–the–counter tablets often contain starch as a filler and exhibit this color reaction. You may have also noticed this dark blue color if you have gotten iodine on a freshly washed and starched shirt!

Starch represents a type of molecule that you have not yet encountered in general chemistry to this point in the course. It is a giant molecule called a polymer or macromolecule that plants produce. Analysis of starch shows that it consists of simple sugar molecules called glucose linked together in specific ways to form a giant, branching chain. A glucose unit contains six carbon atoms and has an empirical formula of CH_2O. This formula looks like *hydrated carbon*, and macromolecules containing simple sugars like glucose are thus called *carbohydrates*.

Procedure

With each household products that you study, identify the active ingredient(s) on the label that is responsible for a particular property. You may have to use your textbook to determine the chemical formula of the compound(s). The household products and the tests to apply will be arranged on the reagent shelf. Use **small** amounts of each product for the tests. Special waste disposal will be noted on the product if required. If not indicated, wash all wastes down the sinks with copious amounts of tap water.

I. Acids and bases

For Solids—Place an amount about the size of a pea in a test tube and dissolve the solid in about 2 mL of water. Dip a glass stirring rod into the solution and then transfer a drop to the test paper, *or* add a few drops of the universal indicator to the solution. A chart is available relating the color of the universal indicator with the acidity of the solution.

For Liquids—Use a glass stirring rod with indicator paper or a few drops of universal indicator as noted above.

II. Oxidizing agents

Place a small amount of a solid suspected of containing an oxidizing agent in a test tube and dissolve it in about 2 mL of water. If the suspected substance is a liquid, place about 2 mL in a test tube. Add 1 mL of a solution of sodium iodide to the test tube with the substance being tested. If an oxidizing agent is present, iodine (I_2) forms and turns the solution straw–colored. You can obtain additional evidence for the presence of iodine (and hence in this case, the presence of an oxidizing agent) by adding a few drops of starch indicator and looking for the appearance of a dark blue color.

III. Specific tests

A. NH_3 and NH_4^+

Add about 2 mL of sodium hydroxide solution to the substance (solid or liquid) being tested. Identify the ammonia gas released by the color it turns litmus paper and by its odor. Look in your textbook if you cannot recall whether ammonia is acidic or basic.

B. HCO_3^- and CO_3^{2-}

Add dilute hydrochloric acid to small amounts of compounds suspected of containing the hydrogen carbonate or carbonate ion. Bubbling of carbon dioxide indicates their presence.

C. Mg^{2+}

Dissolve the suspected compound in a small amount of water. Add drops of a solution of sodium hydroxide. If the magnesium ion is present, a white precipitate will appear.

D. Starch

When mixed with iodine, starch forms a deep blue color. To test for starch, place a few drops of an iodine solution on the suspected solid or to a solution that may contain dissolved starch and observe the color that appears.

IV. Cleanup

Empty the indicated wastes into the labeled waste containers. If no special waste container is present, wash the contents of the various tests down the drain with copious amounts of tap water. Thoroughly wash all test tubes and glassware that you have used prior to returning them to your desk drawer.

Laboratory Report
Experiment 5
Everyday Chemicals

Name: _____

Date & Time: _____

Lab Partner: _____

I. Acids and bases

Name of Product	Formula of Active Ingredient	Color of Indicator	Acidic (√)	Basic (√)

II. Oxidizing agents

Name of Product	Formula of Active Ingredient	Results of Tests & Conclusions

III. Specific tests

List the following for each specific test: (1) products taken, (2) specific test applied, (3) results of test, and (4) conclusions from test.

A. For NH_3 and NH_4^+

Name of Product	Specific Test Applied	Results of Test	Conclusions

B. HCO_3^- and CO_3^{2-}

Name of Product	Specific Test Applied	Results of Test	Conclusions

C. Mg^{2+}

Name of Product	Specific Test Applied	Results of Test	Conclusions

D. Starch

Name of Product	Specific Test Applied	Results of Test	Conclusions

Acid–Base Titrations

Purpose

The general objective of this experiment is to apply your knowledge of acid–base chemistry and the stoichiometry of neutralization reactions to analyze commercial products for acid content. You will use volumetric analytical methods to determine the percent acetic acid in commercial vinegar and the concentration of citric acid in lemon juice. Your instructor will indicate whether you are to perform the experiment using the standard titrimetric procedure or computer–interface systems.

Equipment Needed

Standard titrimetric procedure: two 25–mL burets on a buret stand, two 125–mL flasks, beakers (50–mL, 150–mL and 250–mL), 10–mL and 100–mL graduated cylinders, funnel, dropper, stirring rod, ring stand with iron ring, filter paper

Computer–interface procedure: above equipment plus a magnetic stirrer with teflon–coated stir bar, computer interface with pH combination electrode and drop counter, computer

Reagents Needed

commercial vinegar, lemon juice, standardized sodium hydroxide, phenolphthalein and thymolphthalein indicators in dropper bottles for standard titrations, deionized water

Discussion

When acids and bases react in aqueous solution, the products typically formed are a salt and water as shown below. This type of reaction is called a neutralization. You may want to review acids, bases, salts and reaction types in your textbook or in the discussion of *Chemical Reactions* (experiment 4) in this manual.

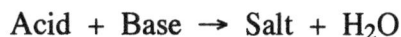

$$\text{Acid } + \text{ Base } \rightarrow \text{ Salt } + \text{ H}_2\text{O}$$

A neutralization reaction can be carried out by adding measured volumes of one reactant (such as the base) to a flask containing an aqueous solution of the other reactant (such as the acid), and this procedure is called a *titration*. When stoichiometric equivalent amounts of acid and base have reacted, the reaction is said to be "complete," and you have reached the *equivalence point* of the titration. The key measurement in titrations is the volume of reactant added to reach the equivalence point, and several techniques are available to determine this volume.

In standard titrations, the device for measuring volumes is called a *buret*. A few drops of an indicator solution are added at the start of the titration. Indicators are compounds that change color when going from acidic to basic solution, and the change is sufficiently intense that only a fraction of a drop of excess acid or base changes the color of the solution. The volume of added titrant that causes the solution to change color is thus the volume needed to reach the equivalence point. The table below lists the indicators and their color changes that you will use in this experiment.

Indicators for Acid–Base Reactions

Indicator	Acidic	Basic
Phenolphthalein	Colorless	Pink
Thymolphthalein	Colorless	Blue

In the computer–interface system, volumes of added reactants are determined by automatically counting drops from a buret and combined with the independently determined volume of each drop. As the titrant is being added, the acidity of the solution is continuously monitored by measuring the pH of the solution. The quantity called "pH" is defined as the *negative of the logarithm of the molar concentration of hydrogen ions.*

$$pH = - \log [H^+]$$

Water is considered neutral and has a pH value of 7.00. Acidic solutions have an excess of hydrogen ions compared to pure water and have pH values below 7. Basic solutions have fewer hydrogen ions than pure water and have pH values above 7.

Details of calculating pH and hydrogen–ion concentrations are beyond the scope of this introduction but can be found in your textbook. From the above equation, however, you should use your calculator and determine that the hydrogen–ion concentration of pure water is 1×10^{-7} M. In acidic solutions, the hydrogen–ion concentration is greater than this value, whereas in basic solutions, it is less.

The practical range of pH measurements is 1 to 14, and thus hydrogen–ion concentrations of 0.1 M to 1×10^{-14} M can be determined with considerable accuracy. Measurement of pH is based on the electrical properties of solutions. A probe, called a combination electrode, is placed in an aqueous solution, and a small potential is generated. A pH meter (or in this case, an interface and computer) measures this potential in millivolts and converts the millivolt values to pH.

A setup for this type of data acquisition consists of the combination electrode connected to an interface which in turn is linked to a computer. Programmable software on the computer directs the collection of titration data into a spreadsheet and also indicates pH values on the computer screen as titrant is being added.

In acid–base titrations, the change in pH at the equivalence point is dramatic over a small addition of titrant. The figure below represents the titration of an acid such as acetic acid with sodium hydroxide, and the equivalence point of the titration occurs at the volume of sodium hydroxide required to reach the inflection point of the pH curve. Computer analysis of your data produces this titration curve and allows you to calculate the volume of sodium hydroxide used to reach the equivalence point.

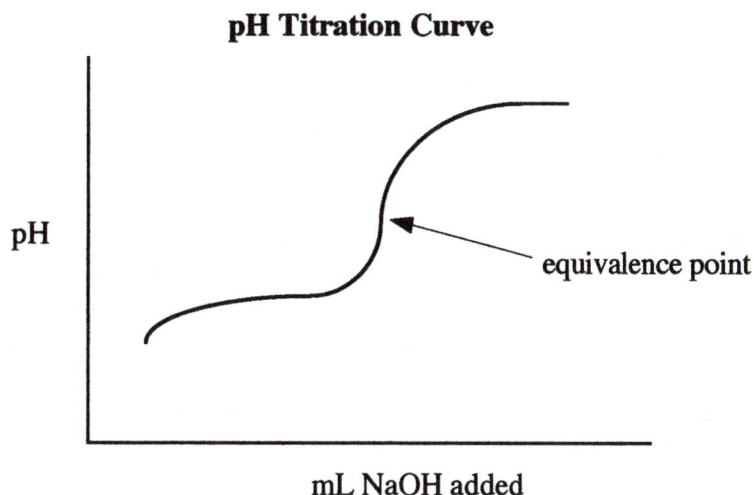

pH Titration Curve

pH

equivalence point

mL NaOH added

In this experiment, you will perform two neutralizations with a solution of known concentration of sodium hydroxide. Solutions of known concentration are called *standardized solutions*. In the first titration, sodium hydroxide reacts with acetic acid in vinegar, and in the second titration, sodium hydroxide neutralizes citric acid in lemon juice.

Acetic acid ($HC_2H_3O_2$) is a monoprotic acid, and citric acid ($H_3C_6H_5O_7$) is a triprotic acid. (Look up the meaning of "protic" if you are not familiar with the term.) When acetic acid is neutralized, the salt formed is sodium acetate ($NaC_2H_3O_2$), and when citric acid is neutralized, sodium citrate ($Na_3C_6H_5O_7$) is produced. The stoichiometry of these neutralizations is shown below.

Vineyar is a solution of acetic Acid + H_2O

Acetic Acid Vinegar

$$HC_2H_3O_2(aq) + NaOH(aq) \rightarrow NaC_2H_3O_2(aq) + H_2O(\ell)$$

Lemon Juice =

$$H_3C_6H_5O_7(aq) + 3\ NaOH(aq) \rightarrow Na_3C_6H_5O_7(aq) + 3\ H_2O(\ell)$$

Citric Acid + H_2O   *Citric Acid*   *Sodium Citrate*

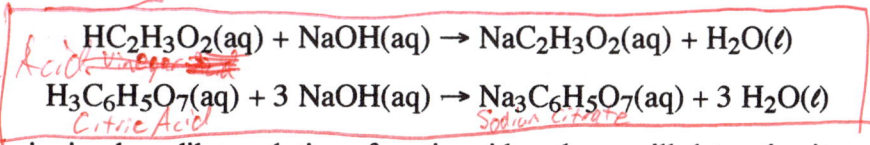

Vinegar is simply a dilute solution of acetic acid, and you will determine its concentration in percent acetic acid by mass. Citric acid is the major acid in lemon juice and is responsible for the sour taste of lemons. You will calculate its concentration in terms of grams of citric acid per liter of lemon juice.

Calculations in neutralization reactions are based on the stoichiometry of the balanced chemical equation. The moles of base used to reach the end point are calculated from the relationship below. This equation simply reflects a rearrangement of the definition of molarity.

Lemon Juice

(2 mL)

$$\text{Volume} \times \text{Concentration} = \text{Amount}$$

(in L)　　　(in M)　　　(in mol)

The volume in liters of NaOH used to reach the equivalence point times the given molarity of NaOH yields the number of moles of NaOH used in the reaction. Combining the number of moles of base with the molar ratio of acid to base in the balanced equation gives the number of moles of acid present in the reaction. The number of moles of acid present divided by the volume of acid used in liters yields the molarity of the acid.

Further calculations require consideration of dilution factors, molar masses of the acid present in the titration and your knowledge of stoichiometry.

Procedure

I. Determination of percent acetic acid in vinegar by the standard procedure

A. Dilution of vinegar

Obtain about 15 mL of vinegar in a clean, dry 50–mL beaker, and add exactly 10.00 mL of this sample to a 10–mL graduated cylinder. You may have to use a dropper to get this exact amount.

Fill a 100–mL graduated cylinder with about 50 mL of deionized water, and then add the accurately measured 10 mL of vinegar from the smaller cylinder. The marking "TC" on the small cylinder means that the cylinder *contains* the indicated amount. Thus you need to rinse the small cylinder with several mL of deionized water and add the washings to the large cylinder.

Dilute the contents of the large cylinder to exactly 100 mL with more deionized water. Empty the cylinder into a clean, dry 250–mL beaker and stir the solution well with a stirring rod. This solution is the *diluted vinegar solution*.

B. Titration of the acetic acid in vinegar

Fill one buret with the *diluted vinegar solution* prepared above. Fill a second buret with standardized NaOH from the stock solution. Make certain that the tips of both burets contain no air bubbles.

Take the initial volume reading from both burets. You do not have to be exactly on the 0.00–mL mark but need to be between 0 and 3 mL. Remember to *record all buret readings to the nearest 0.01 mL!*

Drain about 20 mL of the diluted vinegar from one buret into a clean 125–mL flask. Add 2 drops of phenolphthalein solution.

Titrate with standardized NaOH until one drop turns the entire solution to a very light pink color. Develop your technique so that you can manipulate the stopcock of the buret with one hand while swirling the flask with the other. If you accidentally add too much NaOH,

Vinegar

0.51 mL

NaOH

.30 mL

NaOH

18.95 mL

−.30 mL

NaOH

18.65 mL

– 40 –

you can add a few more drops of vinegar from its buret, and the solution will return to colorless. When one drop of base changes the color of the solution to light pink, record the final volume readings of both burets.

Repeat this titration.

C. Calculation of the percent by mass of acetic acid in vinegar

From the volumes of diluted vinegar and standardized NaOH used and the given molarity of the NaOH, calculate the molarity of the acetic acid in the diluted vinegar.

The sample of vinegar that you titrated was diluted 10 times (10 mL were diluted to 100 mL). Thus multiply the calculated molarity of acetic acid in the diluted vinegar by 10 to obtain the concentration of acetic acid in the straight vinegar.

Convert the moles of acetic acid per liter of vinegar (which is the last molarity calculated above) to grams of acetic acid per liter of vinegar using the molar mass of acetic acid.

Determine the mass of one liter of vinegar from its density of 1.01 g/mL.

Lastly calculate the percent by mass of acetic acid in vinegar, and determine the average of this value for your two trials.

II. Determination of the grams of citric acid per liter of lemon juice by the standard procedure

A. Filter about 5 mL of lemon juice into a small beaker or test tube. Accurately measure 2 mL of the filtered juice with a 10–mL graduated cylinder and add this amount to a 125–mL flask. Remember to rinse the cylinder with a few mL of deionized water and combine the washings with the sample. Add another 20 mL of deionized water and 2 drops of thymolphthalein to the flask. The color of the solution should be pale yellow.

B. Titration of citric acid to the thymolphthalein end point

Fill the buret containing base with more standardized NaOH, and take an initial volume reading. Titrate the lemon juice with base until the solution turns a definite blue–green color, and record this volume reading. This color change is not as sharp as in the previous titration, and you may have to take several readings of the base buret as you notice changes in the color of the solution.

Repeat this titration.

C. Calculation of grams per liter of citric acid in lemon juice

First calculate the moles of sodium hydroxide used to titrate the acid produced by citric acid in lemon juice.

Convert the moles of sodium hydroxide used into moles of citric acid present remembering that citric acid is a triprotic acid. Dividing this number of moles of citric acid by the volume of lemon juice taken for the titration gives the molarity of citric acid in lemon juice.

Finally, convert the molarity of citric acid into grams per liter using the molar mass of citric acid as the conversion factor.

III. Cleanup

Empty all solutions into the waste container provided.

Rinse both burets thoroughly with tap water. Then invert the burets on the stand and open the stopcocks.

Clean all glassware with soapy water, rinse with tap water, leave the burets at your work station.

IV. Analysis of vinegar and lemon juice using a computer–interface system

Review the discussion in Appendix A of this manual on the use of computer–interface systems in chemical experimentation.

Prepare the diluted sample of vinegar and the filtered sample of lemon juice as described for the standard titrimetric procedure given above.

Your instructor will provide you with detailed instructions for setting up the computer and interface as well as specific directions for using the data–acquisition and analysis software program on the computer. General instructions for this process are given below.

Plug the combination pH electrode and drop counter into the appropriate jacks on the interface. Fill one buret with standardized NaOH and position it over the drop counter. Open the data–acquisition software as directed.

Calibrate the pH electrode with standard pH 7.00 buffer that will be provided.

Design the experimental procedure for collection of data by selecting a previously written program (or designing one yourself) as indicated by your instructor.

For the determination of **acetic acid in vinegar**, add 20 mL of diluted vinegar (accurately measured to ±0.01 mL from a buret) to a 150–mL beaker. Place the beaker on a magnetic stirrer, add a teflon–coated magnetic stir bar, and position the pH electrode in the beaker so that it is immersed but not hit by the spinning stir bar.

For the analysis of **lemon juice**, add 2–3 mL (accurately measured from a 10–mL graduated cylinder) to a 150–mL beaker. Add about 25 mL of deionized water and arrange the beaker with pH electrode and stir bar on the magnetic stirrer as described above.

Record the volume on the NaOH buret **prior** to starting a titration.

Acquire data by following instructions as directed by the software. Continue data acquisition until the pH reaches about 10–12 and levels off for several mL. Save your data in a file as instructed, and record the **final** buret reading. From the total amount of NaOH added and the total number of drops analyzed, determine the volume of NaOH per drop.

Analyze your data by opening the file that you have just saved and following the directions of your instructor. Remember that your objective is to determine as accurately as possible the volume of NaOH required to reach the equivalence point. Construct a graph from your data in which you plot pH on the *y*–axis and volume of NaOH in mL on the *x*–axis and identify the equivalence point. A first–derivative graph of these data may help you identify this volume.

As in the standard titrimetric procedure, each analysis should be done in duplicate.

After you have determined the volume of NaOH required the reach the equivalence point in each analysis, calculate the percent acetic acid in vinegar and grams of citric acid in lemon juice as described for the standard procedure.

Lastly, follow the same cleanup and waste–disposal procedures described above and leave your work station in order.

Determination of the Gas Constant

Purpose

The objective of this experiment is to determine the value of R, the gas constant. You will use the ideal gas equation and Dalton's law of partial pressures in this determination.

Equipment Needed

burner and sparker, ring stand, test tube, test tube clamp, pinch clamp, 250–mL beaker, 100–mL graduated cylinder, 250–mL flask or bottle, two–hole stopper fitted with glass tubing, one–hole stopper fitted with glass tubing, rubber and glass tubing, thermometer, rubber bulb

Regents Needed

potassium chlorate, manganese(IV) oxide, water

Discussion

At moderate pressures and temperatures, most gases exhibit *ideal* behavior, and under these circumstances, the quantities of pressure (P), volume (V), temperature (T) and number of moles (n) of the gas are related by what is called the *ideal gas equation*. This equation is

$$P V = n R T$$

The symbol R represents the *gas constant*. Chemists commonly use the following units for these quantities: P in atmospheres (atm), V in liters (L), n in moles (mol) and T in degrees kelvin (K), and thus the units of R are liter atmospheres per mole per degree kelvin ($L \cdot atm \cdot mol^{-1} \cdot K^{-1}$). The commonly accepted value of R is 0.08206 $L \cdot atm \cdot mol^{-1} \cdot K^{-1}$.

In this experiment, you will generate oxygen gas by the thermal decomposition of potassium chlorate and verify the value of the gas constant to three significant figures by determining the pressure, volume, temperature and moles of the oxygen gas produced. The reaction is

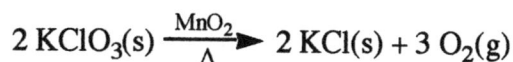

$$2\ KClO_3(s) \xrightarrow[\Delta]{MnO_2} 2\ KCl(s) + 3\ O_2(g)$$

Manganese(IV) oxide (MnO_2) acts as a catalyst that promotes the steady decomposition of potassium chlorate at about 400°C. The delta symbol (Δ) represents heat that you will supply with a flame from a burner. In this reaction, all reactants and products except O_2 are solids, and the heat drives the oxygen gas from the reaction tube. Thus changes in the mass of the reaction tube represent the mass of oxygen produced. From this mass of oxygen, you calculate the moles of oxygen (n) formed.

Oxygen is collected over water in the arrangement shown below. As the gas is produced, it displaces an equal volume of water into a beaker. The volume of the displaced water thus equals V, the volume of the gas.

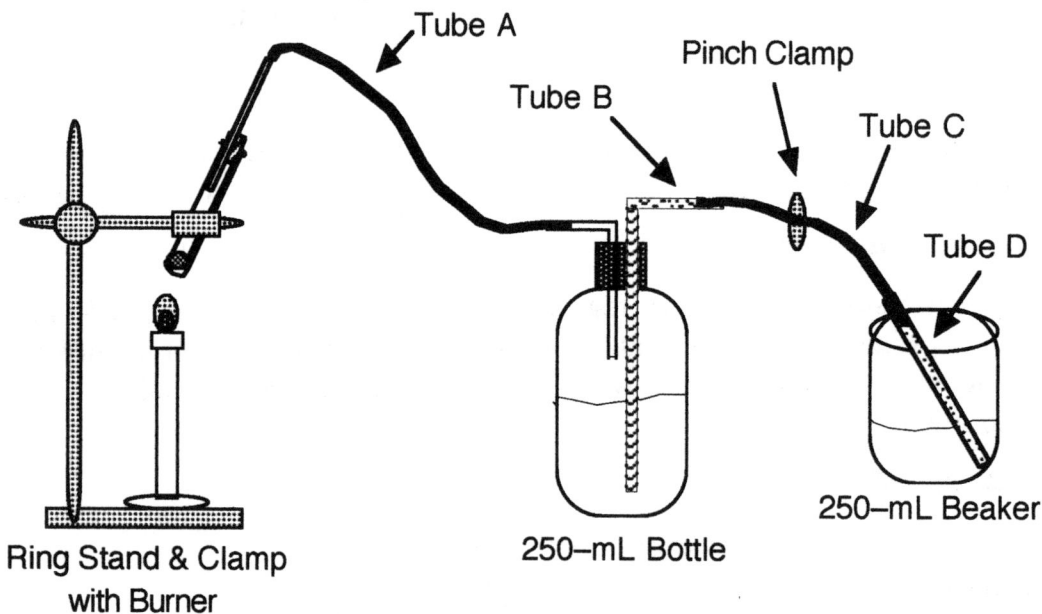

Apparatus for Determination of the Gas Constant

Because the gas is collected over water, water vapor mixes with the oxygen, and both gases contribute to the total pressure of the system. According to Dalton's law of partial pressures, the total pressure of the system equals the sum of the partial pressures of oxygen and water vapor. A partial pressure is just the pressure that a gas would exert if it were there alone. The apparatus is arranged so that the total pressure of the two gases approximately equals the atmospheric pressure, and the equation below represents this relationship.

$$P_{atm} = P_{H_2O} + P_{O_2}$$

The atmospheric pressure (P_{atm}) is easily determined from a barometer. The partial pressure of water (P_{H_2O}) is called the *vapor pressure* of water, and tables are available that list its value at various temperatures. The partial pressure of oxygen (P_{O_2}) is calculated by subtraction and is the pressure P of the oxygen produced in the reaction.

Finally, the temperature T of the gas is the same as that of the displaced water which is measured with a thermometer. From these measurements of *n*, V, P and T expressed in acceptable units, you calculate an experimental value of R.

Procedure

I. Data Collection

Assemble the apparatus as shown on the previous page, and attach a pinch clamp to tube C. The equipment must be clean. Use caution with the glass tubing as it is easily broken and can cut you. Most necessary tubing and fittings are already prepared. The test tube must be clean and *dry*.

Remove the test tube from the apparatus. Fill the 250–mL bottle with water, but do not allow the water to reach tube A. Replace the stopper in the 250–mL bottle. Release the pinch clamp on tube C, and with a rubber bulb force air through the free end of tube A until the tubing system with tubes B, C and D is filled with water. Some water should now be in the beaker. Raise the beaker with tube D below the water surface, and siphon water back into the bottle until the water level is almost at the top of the bottle but not in tube A. Attach and close the pinch clamp on the rubber tube C.

In a plastic weigh boat, weigh about 3 g of potassium chlorate. In the same boat, weigh about 0.1 g of manganese(IV) oxide. Add the solid mixture to the clean, *dry* test tube. Gently tap the tube on the desk to shake all the solid to the bottom of the tube. Weigh the test tube and its contents on the balance to the nearest 0.001 g. Reconnect the test tube with the potassium chlorate–manganese(IV) oxide mixture to the apparatus.

Release the pinch clamp on tube C. If the water levels change, there is a leak in the system. If so, check all connections and repeat the steps in the second paragraph above. When the system has no leaks with the clamp released, close the pinch clamp again. Empty and dry the beaker. Then place the beaker back under tube D. Release the pinch clamp again. No more than a couple of drops of water should flow into the beaker. If more water continues to run into the beaker, recheck the system for leaks once more.

With the pinch clamp released, gently heat the test tube and its contents until about 150 to 200 mL of water has been displaced into the beaker. You can control the rate of oxygen production by altering the rate of heating with the flame. Stop the reaction before oxygen enters tube B. Also do not allow the molten mixture to contact the rubber stopper; an explosion might result. Allow the system to return to room temperature with the clamp open. Keep tube D below the surface of the water in the beaker at all times as some water will return to the flask as the produced oxygen cools.

After the system has cooled, close the pinch clamp at C. Measure the volume of water displaced into the beaker to the nearest 0.1 mL in a 100–mL graduated cylinder. You will have to make two measurements as more than 100 mL of water is displaced.

Disassemble the apparatus and measure the temperature of the water in the flask to the nearest 0.1 degree. Remove the test tube and weigh it to the nearest 0.001 g.

Repeat the procedure. You have enough potassium chlorate in the test tube for the second run and will not have to weigh out any additional compound. Simply use the mass from the final determination in the first run as the initial mass for the second run. Do not attempt to match the volumes of water displaced in the two runs.

Obtain the vapor pressure of H_2O at the determined temperature from the table at the end of this procedure, and then measure atmospheric pressure on the barometer in the laboratory. You may have to interpolate values in the table to find the vapor pressure of water at your measured temperature.

II. Calculations

The mass of oxygen formed represents the difference in masses of the test tube before and after the reaction. A mole of oxygen weighs 32.00 g, and thus you can calculate the number of moles of oxygen produced. This value is n.

The pressure of oxygen is determined by subtracting the vapor pressure of water at the temperature of the flask from the barometric pressure. Convert the units into atmospheres (1 atm equals 760 torr or 760 mm Hg). This value is P_{O_2}.

The temperature T is taken from the temperature of the water in the flask. Convert the reading in degrees celsius to degrees kelvin (°C to K). The volume of the gas is the volume of water that was displaced into the beaker. Change the units into liters, and this value represents V.

From the calculated values of n, P, T and V, calculate the value of R. Determine the percent error of your work compared to the accepted value of 0.0821 $L \cdot atm \cdot mol^{-1} \cdot K^{-1}$ (to three significant figures).

III. Cleanup

After the test tube has cooled, add a few mL of tap water, and loosen the contents with a test tube brush. Empty the contents of the test tube into a labeled waste receptacle. Disassemble the apparatus and clean up your work area.

Vapor Pressure of Water

°C	torr	°C	torr	°C	torr	°C	torr	°C	torr
15.5	13.2	18.5	16.0	21.5	19.2	24.5	23.1	27.5	27.5
16.0	13.6	19.0	16.5	22.0	19.8	25.0	23.8	28.0	28.3
16.5	14.1	19.5	17.0	22.5	20.4	25.5	24.5	28.5	29.2
17.0	14.5	20.0	17.5	23.0	21.1	26.0	25.2	29.0	30.0
17.5	15.0	20.5	18.1	23.5	21.7	26.5	26.0	29.5	39.9
18.0	15.5	21.0	18.6	24.0	22.4	27.0	26.7	30.0	31.8

1 torr = 1 mm Hg

Laboratory Report
Experiment 7
Determination of the Gas Constant

Name: _____

Date & Time: _____

Lab Partner: _____

I. Measurements	Trial 1	Trial 2
Initial mass of tube and contents in g		
Final mass of tube and contents in g		
Volume of H_2O displaced in mL		
Temperature of H_2O in °C		
Vapor pressure of H_2O in torr		
Barometric pressure in torr		
II. Calculations		
Mass of oxygen evolved in g		
Moles of oxygen evolved (n)		
Volume of oxygen in L (V)		
Temperature in K (T)		
Pressure of dry oxygen in torr		
Pressure of dry oxygen in atm (P)		
Calculated value of gas constant R		
Average R value		
Percent error		

Enthalpy of Formation of Magnesium Oxide

Purpose

The objective of this experiment is to determine the enthalpy of formation of magnesium oxide by measuring the enthalpy changes for two reactions by calorimetry and applying Hess's law to the collected data and calculations.

Equipment Needed

Standard procedure: coffee–cup calorimeter (two styrofoam cups nestled inside each other) and lid, thermometer, glass stirring rod, 100–mL graduated cylinder

Computer–interface procedure: coffee–cup calorimeter with lid, magnetic stirrer with teflon–coated stir bar, computer interface with temperature probe, computer

Reagents Needed

solution of 1 M hydrochloric acid, magnesium ribbon, magnesium oxide

Discussion

Enthalpy is one of several quantities, called state functions, that describe a system. Enthalpy is often referred to as the *heat content* of a system and is given the symbol H. Although absolute values of enthalpy cannot be measured, a change in enthalpy between reactants and products or between two states can be determined. This change in enthalpy, ΔH, equals the heat flow in the process at constant pressure as depicted by the equation below in which q_P represents the heat flow under constant–pressure conditions.

$$\Delta H = q_P$$

Calorimetry measures heat flow, and measurements at constant pressure lead to the determination of enthalpy changes according to the above equation. Constant–pressure conditions are easily obtained in a calorimeter that has a small opening to the atmosphere. Atmospheric pressure remains relatively constant over the duration of a short experiment and buffers small changes in pressure that may occur in the calorimeter. Styrofoam coffee cups with lids that do not fit too tightly are amazingly effective constant–pressure calorimeters and are used in this experiment.

Calorimeters are isolated systems in which neither energy nor mass is exchanged with the surroundings. A complete system consists of three parts—a reaction, the surrounding solution and the walls of the calorimeter. The fundamental relationship of calorimetry states that energy as heat is simply transferred among these three components of the system (*i.e.*, the reaction itself, the solution in which the reactions occur and the walls of the calorimeter) but not out of the calorimeter. In equations, these concepts are represented as

$$q_{sys} = 0 \qquad \text{(no heat leaves the calorimeter)}$$

$$q_{sys} = q_{rxn} + q_{soln} + q_{cal}$$

The subscripts refer to the heat flow within the system (*sys*), reaction (*rxn*), surrounding solution (*soln*) and walls of the calorimeter (*cal*).

The part of the system of interest is the heat of reaction, q_{rxn}, and solving the above equations for q_{rxn} gives the following relationship

$$q_{rxn} = -(q_{soln} + q_{cal})$$

Often the heat that flows into or out of the walls of the calorimeter, q_{cal}, is small in comparison to the total heat flow and can be neglected. Such a situation applies in this experiment, and the last equation then reduces to

$$q_{rxn} = -q_{soln}$$

and the heat flow of the reaction is opposite in sign to q_{soln}, the heat flow into or out of the solution that surrounds the reaction. In an *exothermic* reaction, heat flows out of the reaction into the surrounding solution, and the heat q is given a negative sign. For an *endothermic* reaction, heat flows from the surrounding solution into the reaction, and the sign for q is positive.

The figure below depicts a calorimeter with the three components of the system discussed above. The reaction is shown with a discreet boundary, but in reality the reaction is uniformly distributed throughout the solution. A hypothetical boundary is drawn in this model for clarity in thinking about heat flow in the system. Heat (q) is flowing from the reaction into the solution, and thus the reaction depicted is exothermic.

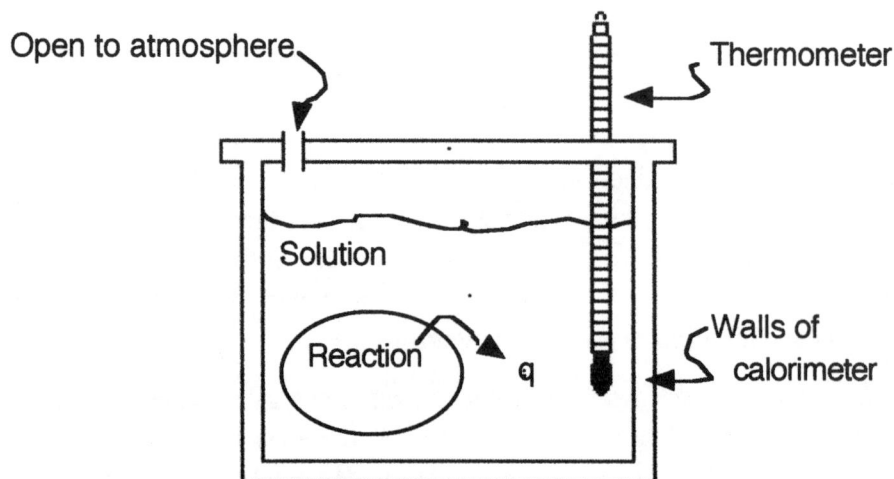

Constant–Pressure Calorimeter

In a calorimeter, temperature measurements are made in the solution surrounding the reaction, and the heat flow that is measured is q_{soln}. Knowledge of the specific heat capacity (s) of the solution, its mass (m) in grams and the difference in final and initial temperatures, (ΔT), leads to the equation for calculating the heat flow into or out of the solution.

$$q_{soln} = (s_{soln})(m_{soln})(\Delta T_{soln})$$

The units for specific heat capacity are $J \cdot g^{-1} \cdot {}^{\circ}C^{-1}$. When s is multiplied by the mass in g and ΔT in $^{\circ}C$, the product q is in J.

The first step in a calorimetry experiment is to determine q_{soln}. Then changing the sign of q_{soln} gives q_{rxn}. These values represent heat flow in J for the *given quantity of the reacting substance*. Determining ΔH requires calculating q_{rxn} for a *mole* of the reacting species. Conversion from grams to moles requires the molar mass of the reactant. Your textbook discusses specific examples of these calculations.

In this experiment, you will determine ΔH for two reactions by calorimetry and will use the *accepted value of ΔH for a third reaction*. These reactions are

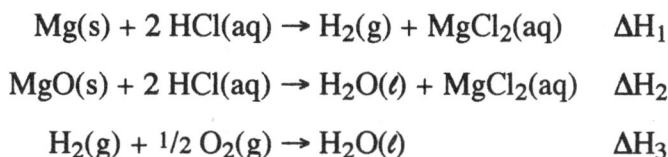

$$Mg(s) + 2\,HCl(aq) \rightarrow H_2(g) + MgCl_2(aq) \qquad \Delta H_1$$

$$MgO(s) + 2\,HCl(aq) \rightarrow H_2O(\ell) + MgCl_2(aq) \qquad \Delta H_2$$

$$H_2(g) + 1/2\,O_2(g) \rightarrow H_2O(\ell) \qquad \Delta H_3$$

These data (ΔH_1, ΔH_2, ΔH_3) are then combined to calculate the enthalpy change for the formation of magnesium oxide. A *formation reaction* is one in which one mole of a pure substance is formed from its pure elements in their most stable form. Thus for magnesium oxide, the formation reaction is

$$Mg(s) + 1/2\,O_2(g) \rightarrow MgO(s) \qquad \Delta H_f$$

The symbol for the enthalpy of formation includes the subscript *f* as shown above ΔH_f.

The thermochemical relationship that allows you to combine reactions and ΔH values is called Hess's law. This law states that if you can add a set of reactions to give an overall reaction, then ΔH for the overall reaction is simply the sum of the ΔH values for each reaction in the set. Again, your textbook discusses Hess's law in some detail.

You might note that enthalpy data for formation reactions are usually given under standard state conditions represented by the superscript " $^{\circ}$ " in the term ΔH_f°. Standard state conditions for a substance are the most stable form of the substance at 1 atm pressure and a specified temperature, which is usually 25°C. In this experiment, you will be working close enough to standard state conditions at 25°C and 1 atm so that you can use the tabular values in your textbook for the standard enthalpy of formation of magnesium oxide to compare your results.

Procedure

I. Standard procedure for calorimetry

A. Reaction with magnesium

Weigh a dry calorimeter (without its cover) to the nearest 0.001 g. Add 100 mL of 1 M HCl(aq) from a graduated cylinder, and reweigh the calorimeter with the solution. This second mass can only be recorded to the nearest 0.01 g.

Obtain a small piece of magnesium ribbon, and weigh approximately 0.24 g. Be sure to record this measurement to the nearest 0.001 g.

Assemble the calorimeter with its lid, thermometer and stirring rod as shown in the figure below.

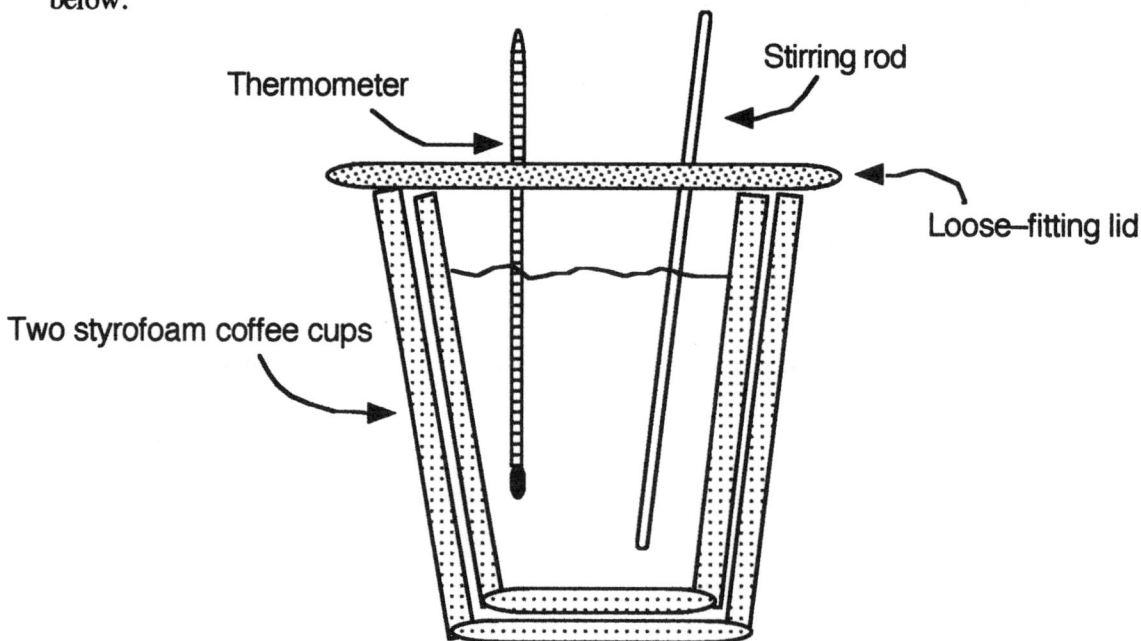

Coffee–Cup Calorimeter

Record the temperature of the hydrochloric acid to the nearest 0.1°C. Remove the cover of the calorimeter and add the magnesium. Then quickly replace the cover. Observe the temperature and gently stir the solution with the glass rod until the temperature stops rising. Record the highest temperature to the nearest 0.1°C.

Empty the calorimeter contents into the waste container provided. Dry the calorimeter with a paper towel.

B. Reaction with magnesium oxide

Repeat the above procedure except use approximately 0.8 g of magnesium oxide (measured to the nearest 0.001 g) instead of magnesium ribbon.

Empty the calorimeter contents and dry the calorimeter.

C. Calculations

Calculate the mass of hydrochloric acid and ΔT for both reactions by subtraction of the appropriate quantities.

From these respective values, calculate q_{soln} for each reaction using a *specific heat capacity for 1 M HCl as 3.64 J·g⁻¹·°C⁻¹*. From q_{soln}, determine q_{rxn} for each reaction. You may assume that q_{cal} in this experiment is negligible.

Finally, calculate q_{rxn} values on a mole basis using the molar masses for magnesium and magnesium oxide, respectively, and convert from J to kJ. *When q_{rxn} is expressed in terms of kJ (or J) per mole at constant pressure, it becomes ΔH.*

Refer to the *Discussion* of this experiment, and write the balanced equations that you will be using and their respective ΔH values. The ΔH value for the third reaction can be found in the tables for standard enthalpy of formation in your textbook.

Rearrange these equations so that when added they will give the formation reaction for magnesium oxide. Follow the laws of thermochemistry and change ΔH values accordingly. Lastly, add the rearranged equations and ΔH values to give the net reaction and ΔH_f for the formation of magnesium oxide.

Look up the accepted value for ΔH_f under standard state conditions in your textbook, and calculate the percent error of your experiment. Be careful in reporting your percent error to the correct number of significant figures. If you are uncertain how to do this calculation properly, look up the rules in the front of this manual!

II. Calorimetry using computer–interface systems

A. Review the discussion in *Appendix A* of this manual on the use of computer–interface systems in chemical experimentation. Your instructor will provide specific instructions for use of the computer interface system in this experiment.

Weigh a dry calorimeter (without its cover) to the nearest 0.001 g. Add 100 mL of 1 M HCl(aq) from a graduated cylinder, and reweigh the calorimeter with the solution. This second mass can only be recorded to the nearest 0.01 g.

Obtain a small piece of magnesium ribbon, and weigh approximately 0.24 g. Be sure to record this measurement to the nearest 0.001 g.

Assemble the calorimeter with its lid, as shown on page 56 but replace the thermometer with the temperature probe attached to the computer–interface and the stirring rod with a teflon–coated stir bar. Arrange the calorimeter on a magnetic stirrer and secure it with an iron ring attached to a ring stand to prevent the calorimeter from accidentally tipping over.

Calibrate the probe using a beaker of ice water containing a thermometer. Follow the instructions on the computer screen. Repeat the process with a beaker of hot tap water for the second calibration point. When the calibration is complete, replace the temperature probe in the calorimeter and adjust the magnetic stirrer so that the solution is stirring gently.

Design the experimental procedure for data collection by selecting a previously written program (or designing one yourself) as indicated by your instructor.

Acquire data for about a minute prior to the addition of the previously weighed piece of magnesium ribbon. Add the magnesium ribbon and continue data collection until the temperature has become constant for about 2 minutes. Save your data in a file as instructed.

Analyze your data from the saved file by graphing temperature on the *y*–axis versus time on the *x*–axis. Use the graphical information to determine *initial* and *final* temperatures and record these values on your report sheet.

Empty the calorimeter contents into the waste container provided. Dry the calorimeter with a paper towel.

Repeat the above procedure except use approximately 0.8 g of magnesium oxide (measured to the nearest 0.001 g) instead of magnesium ribbon. The temperature probe does not need recalibration. You begin this second calorimetry experiment by selecting the same program used in the *Design* paragraph above and following the same steps in the *Acquire* and *Analyze* paragraphs.

Empty the calorimeter contents and dry the calorimeter. Also rinse the probe with deionized water and leave the computer–interface and computer as instructed.

Collection and analysis of data are rapid with the computer–interface system, and if you are uncertain about the initial and final temperatures of the reaction, you will have sufficient time and reagents to repeat the procedure several times.

Enter collected data in the appropriate lines on the *Laboratory Report* and complete the calculations and reactions as indicated.

Absorption Spectroscopy

Purpose

The principle objective of this experiment is to study how scientists and technicians use the absorption of light by solutions to determine solute concentration. Specifically, the aims are to determine an absorption spectra of a solution of potassium permanganate, to establish the relationship between absorption and its concentration, and lastly to determine the potassium permanganate concentration in an unknown solution.

Equipment Needed

Spectronic 20 or other similar visible wavelength spectrophotometer, two cuvets (specially marked test tubes), 4 test tubes, test tube rack, 10–mL and 100–mL graduated cylinders, 150–mL beaker, ruler, sheet of graph paper.

Reagents Needed

stock solution of potassium permanganate, unknown solution, deionized water

Discussion

When electromagnetic radiation impinges on matter, often some energy of the radiation is absorbed. The absorbed wavelengths have energies that correspond to the energies of events at the atomic or molecular level, and these energies, frequencies and associated wavelengths can ultimately be related by the key equation

$$E = h\nu \ \text{ or } \ E = h\,c/\lambda$$

where E represents the energy; ν, the frequency; c, the speed of light; λ, the wavelength of the associated radiation; and h, Planck's constant. For example, atoms within a bond vibrate at frequencies in the range of infrared radiation, and organic chemists identify types of bonding structure by the wavelengths of infrared radiation that a molecule absorbs.

Electromagnetic radiation in the visible part of the spectrum (wavelengths of 400–750 nm) has sufficient energy to move outer electrons to unoccupied orbitals at a slightly higher energy. Compounds of some transition elements with their partly filled *d* orbitals absorb visible light and undergo these electronic transitions. Absorbing part of the spectrum of white light imparts brilliant color to compounds and solutions of these compounds in contrast to the white or gray compounds of metal ions from groups 1A and 2A with their filled *d* orbitals.

Absorption of electromagnetic radiation is a physical property of matter, and instruments that measure absorption as a function of wavelength of light are called spectrophotometers. Absorption spectroscopy refers to analytical procedures based on these principles.

The figure below depicts the principles of a spectrophotometer that operates in the visible region of the spectrum. White light is passed through a prism or set of narrow slits that separates light into wavelengths. Light of a single wavelength (called *monochromatic light*) is then passed through a cell containing a solution of the absorbing compound. Light that passes through the cell then strikes a photocell detector connected to a meter or read–out device. The meter indicates the intensity of the radiation that has passed through the sample.

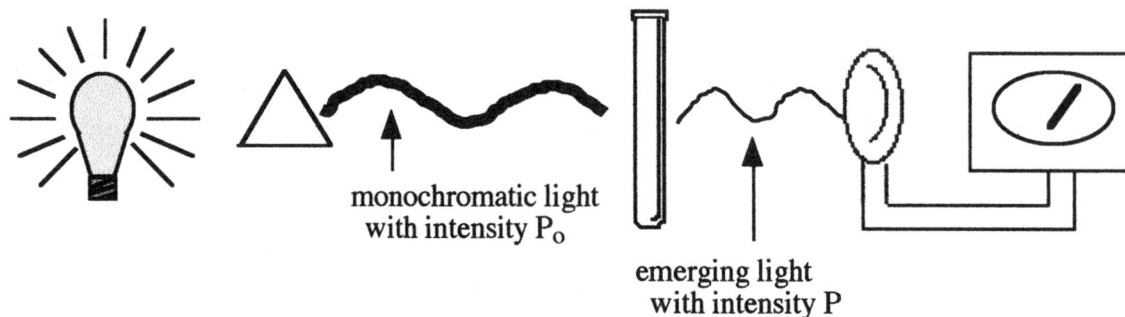

monochromatic light
with intensity P_o

emerging light
with intensity P

The ratio of the intensity of the emerging light P to the intensity of the incident light P_o is called *transmittance T*. Multiplying transmittance by 100 yields percent transmittance.

$$T = \frac{P}{P_o}$$

$$\%T = \frac{P}{P_o} \times 100$$

The meter is calibrated to indicate zero transmittance of light by blocking all light from hitting the detector. Then a blank sample containing all solvent components *except* the sample to be analyzed is placed in the sample compartment, and the meter is adjusted to read 100% transmittance. Then when the sample is placed in the light beam, the spectrophotometer measures only the absorption caused by the sample.

In this experiment, you will examine the absorption properties of the permanganate ion (MnO_4^-) in an aqueous solution of the ionic compound potassium permanganate ($KMnO_4$). The permanganate ion has a distinctive purple color both in the solid state as in $KMnO_4(s)$ and in solution as $MnO_4^-(aq)$. Its purple color results from absorption of light in the green part of the spectrum. In the dissolved state, the MnO_4^- ion absorbs in broad bands covering many wavelengths; only in the gaseous state do atoms, molecules and ions produce sharp line spectra. Potassium ions K^+ do not absorb visible light.

The experiment has three parts. First, you will determine the absorption spectrum of an aqueous solution of $KMnO_4$ from 340–600 nm. The absorbance of a dilute $KMnO_4$ solution will be measured against a blank of pure H_2O at wavelength intervals of 20 nm. A plot of absorbance versus wavelength gives a figure called an absorption spectrum. From this spectrum, you identify the wavelength at which the permanganate ion exhibits maximal absorption.

In the second part of the experiment, you will examine the relationship between absorption and concentration of the absorbing species (which in this case is the permanganate ion). This relationship is called Beer–Lambert's law or simply Beer's law. Verbally, this law says that absorption of monochromatic light is directly proportional to the distance that light travels through an absorbing medium and to the molar concentration of the absorbing species. The mathematical equation that expresses this law is

$$A = a \cdot b \cdot c$$

where A is the absorption; a, an absorptivity coefficient; b, the length that monochromatic light travels through the solution; and c, the molar concentration of the absorbing species. Absorbance is a dimensionless number, and the units of a, b, and c are chosen to cancel completely. Absorbance measurements are usually made at the wavelength of maximal absorption.

The absorptivity coefficient a is just a proportionality constant, and its value depends on the nature of the absorbing species and the wavelength at which the absorbance measurements are taken. The value of a for MnO_4^- at a wavelength at which it absorbs strongly is 2.3 x 10^3 $M^{-1} \cdot cm^{-1}$. This number means that a solution of 1 x 10^{-4} M in permanganate ion would have an absorbance of 0.23 at this wavelength in a 1.0–cm cell. (Try to calculate this value yourself using the expression $A = a \cdot b \cdot c$ and the values given.)

Beer and Lambert's theoretical development of the law also indicates that absorbance (A) is related to transmittance (T) as shown below.

$$A = -\log T$$

Spectrophotometers actually measure transmittance or percent transmittance. Digital instruments convert the signal to absorbance values electronically, whereas analog spectrophotometers (such as some Spectronic 20 instruments which have a meter) provide an accompanying scale. Also the experimenter can calculate absorbances from transmittance values using the equation above.

Values for absorbance A range from 0 to 2.0 on most instruments. Close examination of the meter on an analog instrument (*i.e.*, one on which you determine transmittance or absorbance by reading the position of a needle on a meter) shows that accurate values range only from 0.1 to 0.7 absorbance units. This sensitivity reflects the logarithmic relationship between transmittance and absorbance. Also note that 100% transmittance equals zero absorbance, and 0% transmittance equals infinite absorbance.

$$100\% \, T = 0.000 \, A \quad \text{and} \quad 0\% \, T = \infty \, A$$

In most experiments with Beer's law, you determine absorption in cells with the same dimensions so that the light path stays constant. (The b term in the expression $A = a \cdot b \cdot c$.) Under these conditions, both a and b are constant, and A is directly proportional to the molar concentration c. A plot of A versus c gives a straight line as shown in the figure on the next page. This graph is called a *standard curve* although it is a straight line in theory.

Standard Absorption Curve for MnO4– Ion

absorbance of unknown

$A_{\lambda max}$

0.8
0.6
0.4
0.2

conc. of
unknown

1   2   3   4

Conc. of MnO_4^- (M x 10^{-4})

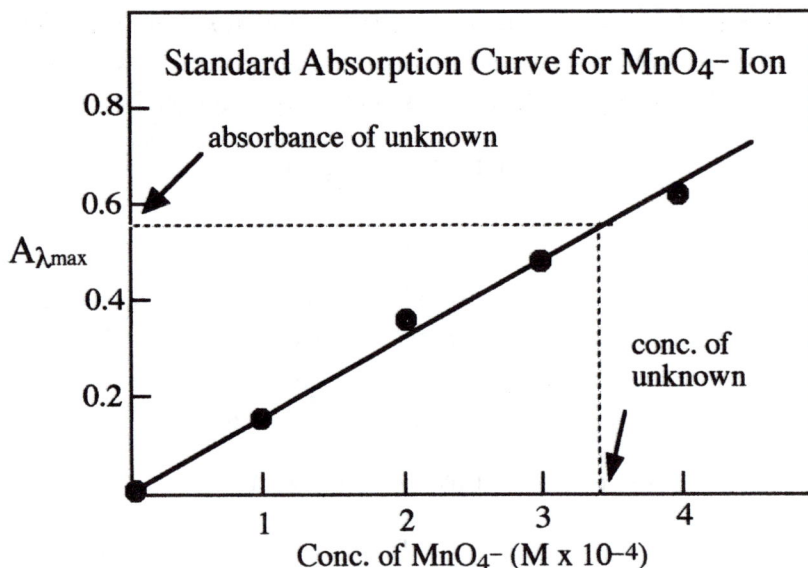

The third part of this experiment consists of determining the concentration of permanganate ion in an unknown sample. The absorbance of a solution containing the permanganate ion is measured at the same wavelength at which the standard curve was determined, and the concentration is read from the standard curve as shown in the figure above.

Techniques described in this experiment can be used to determine the manganese content of various samples. For example, manganese is a common constituent of steel. If time permitted, you could dissolve a small amount of steel in a solution of a strong acid and oxidize the manganese to the permanganate ion. Then you could determine the concentration of permanganate ion in the prepared solution from absorbance measurements and a standard curve just as you will do in this experiment. This procedure, as well as similar ones, is done in many analytical chemistry laboratories.

Procedure

I. Absorption spectrum of a KMnO4 solution

A. Data collection

Turn on the spectrophotometer and allow it to warm up for at least 5 minutes while you are preparing solutions.

Obtain about 20 mL of the concentrated $KMnO_4$ solution in a 50–mL beaker. Record its concentration on the second page of your laboratory report where indicated. Transfer about 5 mL of this solution to a clean cuvet. The height of the liquid in the cuvet should be about 5 cm. This cuvet is called the *sample*. Also please note that the cuvets are special (and expensive) tubes. Be careful with them and do not scratch their surfaces. Cuvets will have a vertical line on them near the top and often have a dot or company logo on them.

Mostly to the top

Fill another clean cuvet with about 5 mL of deionized H_2O. This cuvet is called the *blank*. The volumes for *this* part of the experiment do not have to be exact.

Set the wavelength to 340 nm, and with nothing in the cell compartment, set the readout or needle to 0% T with the dark current control using the left–hand knob on the instrument. You should not have to reset or change this setting during the experiment.

Insert the blank cuvet into the cell compartment, align the vertical line on the tube with the mark on the compartment, close the lid and set the 100% T value with the right-hand knob. *You will have to repeat this calibration to 100% T each time you change the wavelength.*

Insert the sample cuvet into the cell compartment in a similar manner and read absorbance A on the instrument or from the lower scale from the analog instruments that have a meter and needle. When using the analog instruments, remember to estimate values between markings on the scale. Readings should be recorded to three decimal places whenever possible.

An alternate—and more accurate—way to determine absorbance on an analog instrument is to read the %T scale estimated to the nearest 0.1% value. Dividing this reading by 100 converts the measurement to transmittance T, and then taking the negative of the logarithm of transmittance yields absorbance ($A = - log\ T$).

Increase the wavelength 20 nm, and repeat zeroing the instrument with the water blank and measuring absorbance in the sample until you have A readings through 600 nm. Transfer the sample cuvet into a test tube (number 4) for part II of the experiment.

B. Data analysis

Plot A versus wavelength on graph paper with absorbance A on the ordinate (y–axis) and wavelength (λ) in nm on the abscissa (x–axis). You may use the top half of a piece of graph paper for this part. Choose convenient and easily readable units on both axes. The x–axis should start at 320 or 340 nm.

II. Standard curve (Beer's law relationship)

A. Solution preparation

Add to 3 clean test tubes (numbered 1 through 3) the following amounts of $KMnO_4$ and water and mix the contents of the tubes well. Use the 10–mL graduated cylinder for these measurements. Tube 4 should contain a few mL of the stock solution of $KMnO_4$ from part I of this experiment.

Tube	1	2	3	4*
H_2O	7.5 mL	5.0 mL	2.5 mL	—
$KMnO_4$	2.5 mL	5.0 mL	7.5 mL	~5 mL
*Tube 4 contains about 5 mL of the standard $KMnO_4$ solution that you used in part I.				

B. Data collection

Choose the wavelength from your graph where the permanganate solution had the greatest absorption and set the instrument to that value. Set the 100% T value with the blank cuvet that contains H_2O. Save this cuvet to check the zero reading during the experiment. This value also represents zero concentration of permanganate on the standard curve.

Add the contents of tube 1 to a clean cuvet to a height of about 5 cm, insert the cuvet into the cell compartment, and read the absorbance A. Repeat this procedure with tubes 2, 3 and 4. Between readings, rinse the cuvet with a small amount of the contents of the next tube.

C. Data analysis

Calculate the concentration of the standard solutions using the relationship

$$V_1M_1 = V_2M_2$$

Plot absorbance A on the ordinate versus the molar concentration of the standards solutions (in tubes 1 through 4) on the abscissa. Select convenient and easily readable units for both axes. You may use the bottom half of the graph paper for this standard curve. Because H_2O was chosen as the blank and is the solvent in the permanganate solution, the origin is also a data point.

Calculate the best straight line through the data points using a graphing calculator or by following the procedure outlined in *Appendix B* at the end of this manual. The statistical term for *best straight line* is *linear regression* or *least–squares analysis* of the data points. Report the calculated equation in the form of $y = mx + b$ where y is the absorbance A, x is the concentration of the permanganate ion, m is the slope (and thus the value of $a \cdot b$ in the Beer's law expression), and b is the y–intercept. The y–intercept should be close to zero if you have used your calculator to determine the best straight line. Record the equation on the laboratory report where indicated.

III. Analysis of unknown sample

A. Data collection

Leave the spectrophotometer at the same settings that you used for determining the standard curve. Add about 5 mL of the unknown solution to the sample cuvet, and take the absorbance reading as before.

B. Data analysis

Determine the concentration of the unknown from the standard curve that you have calculated and drawn.

IV. Cleanup

Empty all solutions into the waste beaker. Clean and leave the cuvets in a small beaker next to the spectrophotometer.

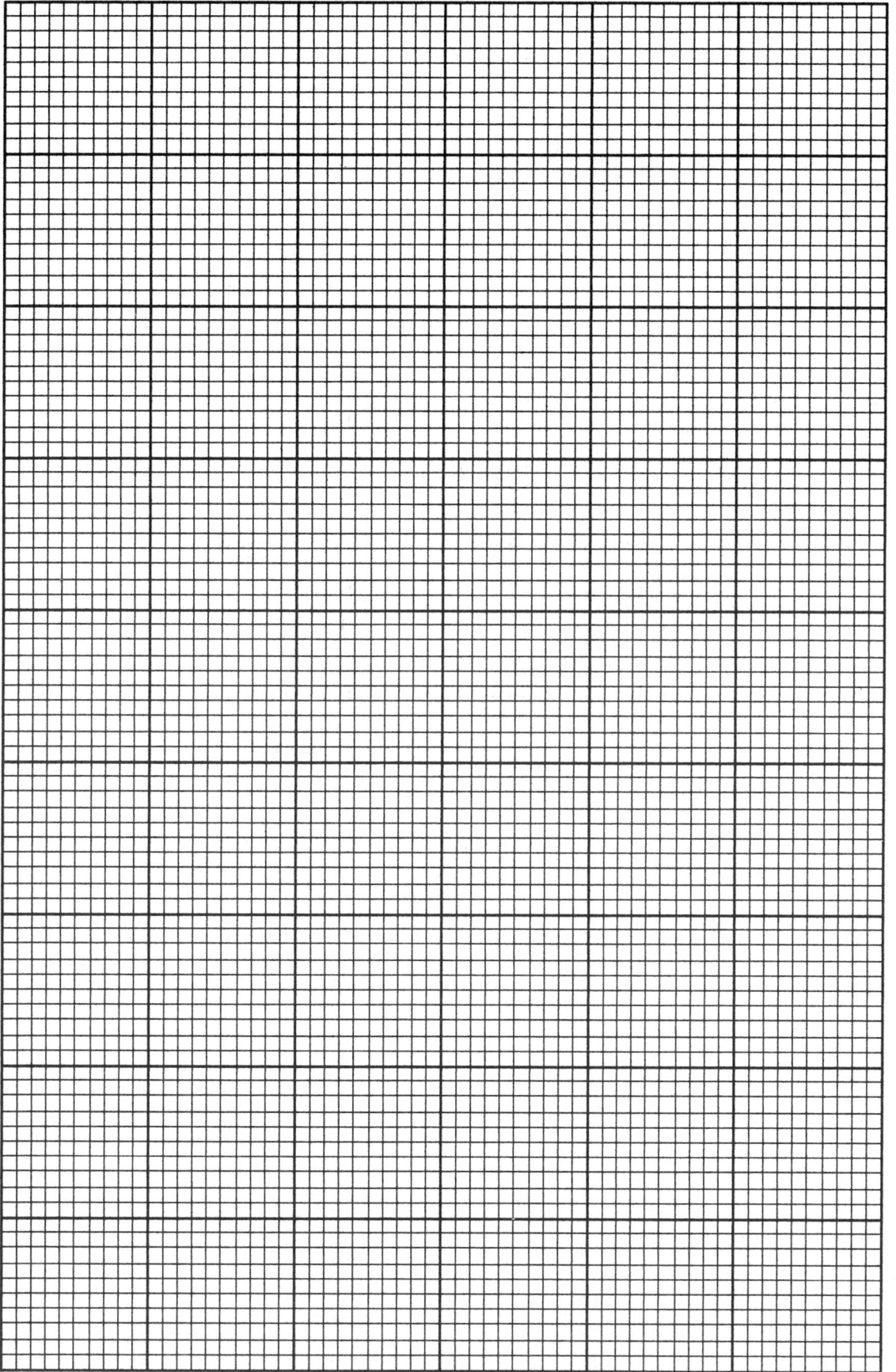

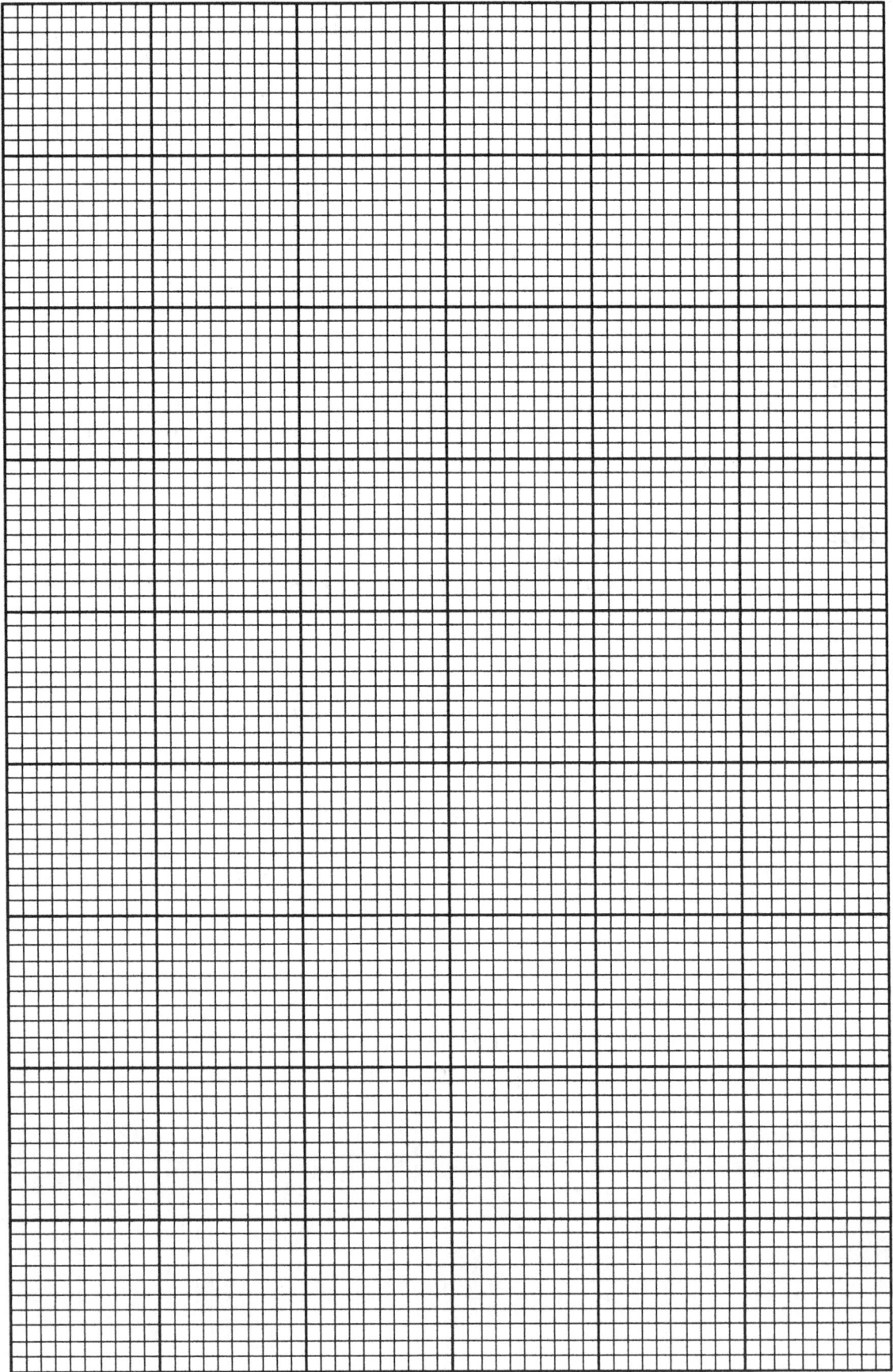

Determination of Phosphorus
in Fertilizer

Purpose

The objective of this experiment is to analyze a sample of fertilizer for its phosphorus content by absorption spectroscopy.

Equipment Needed

spectrophotometer, two cuvets, 10–mL and 100–mL graduated cylinders, six 125–mL Erlenmeyer flasks, three small beakers, weigh boat, spatula, 25–mL buret, 250–mL volumetric flask, two regular Beral pipets, one calibrated 1–mL Beral pipet.

Reagents Needed

ammonium molybdate reagent, tin(II) chloride reagent, standard phosphate solution, deionized water

Discussion

In this experiment, you will use the principles of absorption spectroscopy to determine the amount of phosphorus in a sample of fertilizer. Consult the discussion for the previous experiment to review basic concepts about light absorption and the relationship between absorption and concentration of the absorbing species.

Many fertilizers contain phosphorus in the form of an inorganic salt in which the anion is phosphate (PO_4^{3-}), hydrogen phosphate (HPO_4^{2-}), or dihydrogen phosphate ($H_2PO_4^{-}$). These ions do not absorb light in the visible range. A colored product containing phosphate, however, can be formed with these phosphate ions, and this compound (sometimes called a *complex*) does absorb visible light. Thus you can determine a standard curve that relates the amount of phosphorus in a sample to the degree of absorption of the colored compound. Samples that contain an unknown amount of phosphorus are then treated identically, and the amount of phosphorus in them is obtained from the standard curve.

The colored complex, called molybdenum blue, contains the elements phosphorus, oxygen and molybdenum and absorbs light maximally around 650 nm. Unabsorbed light makes the complex appear blue. Although the precise structure of the blue complex is not known, the reaction that forms it is reproducible, and the concentration of phosphorus is directly proportional to absorption by the complex. Thus you can use this procedure to analyze unknown samples for their phosphorus content.

This experiment consists of two parts. First, you will prepare a standard curve that relates absorption by the complex to the concentration of phosphorus in the complex. To construct this graph, you will prepare four samples of known phosphorus concentration, form the colored complex with each solution, and then determine the absorbance of each sample at 650 nm. Secondly, you will prepare an unknown sample in a similar manner as you did for the standards and determine its phosphorus content using your standard curve.

Fertilizers typically report phosphorus content *as if* all the phosphorus present were in the compound P_2O_5. The percent phosphorus by mass in P_2O_5 is 43.7%. Thus 1.0 mg of phosphorus corresponds to 2.3 mg of P_2O_5. Common fertilizers have 10% phosphorus expressed as P_2O_5.

High levels of phosphate ions cause a serious environmental problem in surface waters such as rivers and lakes. These phosphorus–containing ions promote rapid growth of algae and microorganisms. If the growth continues unchecked, oxygen levels in the water fall, and aquatic life in that body of water dies. Decomposition of the dying aquatic animal and plant species consumes more oxygen exacerbating the problem. This process is called *eutrophication*, and in the 1950's Lake Erie exhibited many properties of a body of water that was dead to aquatic life. Lakes fed by the Chattahoochee River downstream from Atlanta have also shown signs of suffering from excess phosphate in the waters.

Phosphate enters surface waters from the runoff of land after rains and from the use of phosphate–containing detergents that pass through waste treatment facilities. In recent years, many major metropolitan areas, including Atlanta, have banned the use of detergents containing phosphate to eliminate one source of pollution. Fortunately, reducing these human sources of phosphate pollution contributes to the recovery of lakes and rivers from the eutrophication process.

Procedure

I. Standard curve of phosphomolybdate complex

A. Preparation of standard solutions

Fill a 25–mL buret with the standard phosphate solution. Add *exactly* 2.00 mL to a 100–mL graduated cylinder, dilute to 25.0 mL with deionized water, and mix the contents well. Pour this diluted sample into a labeled 125–mL Erlenmeyer flask. Rinse the 100–mL graduated cylinder with deionized water and discard the rinse.

Repeat the above dilution starting with *exactly* 4.00 mL, 6.00 mL, 8.00 mL and 10.00 mL. You will have to refill the buret at least once. Also add 25.0 mL of deionized water to a sixth flask to be used as a blank. At this point you will have 5 flasks containing 25 mL of phosphate solutions and one flask with deionized water.

B. Color development

Carefully add 1.00 mL of the ammonium molybdate reagent to each flask and swirl the flask to mix the contents. Use a calibrated Beral pipet for this addition. Then add 2 drops of the tin(II) chloride reagent from a dropper bottle or using a regular Beral pipet. Again swirl the flask to mix the contents. Note the time of this last addition.

After 5 to 15 minutes from the addition of tin(II) chloride, calibrate the spectrophotometer at 650 nm using the water sample as the blank. If you are using a digital readout spectrophotometer, determine the absorbance of each standard solution at 650 nm. If you are using an analog instrument (such as a Spectronic 20 with meter), record the percent transmittance to the nearest 0.1%. Use your calculator and convert the %T reading to absorbance A to three significant figures.

II. Analysis of fertilizer sample

Weigh about 0.5 g of fertilizer accurately to the nearest 0.001 g. Transfer this sample to a 250–mL volumetric flask, dissolve the sample with deionized water, and dilute to the mark with more water. Cover the flask (with a stopper or Parafilm) and invert it several times to mix the solution thoroughly.

Then dilute 10.0 mL of this solution to 100.0 mL in a graduated cylinder. Empty the contents of the cylinder into a beaker, mix the solution thoroughly, and pour it back into the 100–mL graduated cylinder. This 1:10 diluted solution is the one that you will use in the remainder of this analysis.

At this point, you will have to do some investigation on you own. The purpose is to select a sample size from the diluted sample that will produce an absorbance within the bounds of your data points and preferably between 0.15 and 0.7 absorbance units (or 70 to 20 percent transmittance %T) when the color is developed exactly as you did for the standard phosphate solutions.

For example, start with 1.0 mL carefully measured with a clean, calibrated Beral pipet. Dilute this sample to 25.0 mL with deionized water in the graduated cylinder and mix well. Next transfer this solution to a small beaker or flask. Then to this solution, add 1.0 mL of ammonium molybdate, mix well, and finally add 2 drops of tin(II) chloride. If the color that develops is too dark or too light, repeat the procedure but start with either less sample (<1.0 mL) or more sample (>1.0 mL) in the dilution step. This procedure assures that you treat each aliquot taken for analysis the same as you treated the standard phosphorus solutions.

Once you determine a suitable sample size, repeat the color development on two identical volumes taken from the 1:10 diluted solution in the 100–mL graduated cylinder. You will thus have three identical measurements to average.

III. Data analysis

Calculate the amount of phosphorus in **mg P** in each 25–mL solution used for the standard curve. Multiply the concentration of phosphorus on the standard solution (in mg P/mL) by the volume analyzed. On graph paper, plot absorbance at 650 nm on the y–axis versus the amount of phosphorus in mg P on the x–axis.

Use a graphing calculator or the procedure outlined in *Appendix B* and determine the best straight line through the data points. Record this equation on the laboratory report where indicated.

The amount of phosphorus in the original sample must now be calculated using the sample size, dilution factor and original solution volume. Equations for this process are shown below in which you enter experimentally determined values within the appropriate brackets.

$$\text{mg P in sample} = \left(\frac{[\text{mg P (from graph)}]}{[\text{mL sample analyzed}]} \right) \times \left(250 \text{ mL} \times 10 \right)$$

size of volumetric flask

mL taken from 100–mL graduated cylinder

dilution factor of original sample
(10 mL to 100 mL)

$$\%P = \frac{[\text{mg P in sample}]}{[\text{mg sample}]} \times 100\%$$

In the first equation above, the quotient in the first parenthesis represents the concentration of phosphorus (in mg P/mL) in the 100–mL graduated cylinder. The product in the second parenthesis accounts for the volume into which the dry sample was dissolved and the dilution factor of this original sample solution.

IV. Cleanup

After you have finished all measurements, discard *all* solutions into the waste container. Rinse the buret with deionized water and invert on the stand to dry. Also rinse the cuvets with deionized water and *leave the cuvets next to the spectrophotometers*. Do not put the cuvets into your desk drawers.

Laboratory Report
Experiment 10
Determination of
Phosphorus in Fertilizer

Name: _____

Date & Time: _____

Lab Partner: _____

I.   Standard curve of phosphomolybdate complex

A. Concentration of standard phosphorus solution:
   (on reagent bottle or given by your instructor)   _____ mg P/mL

B. Absorbance readings and calculated amounts of phosphorus (mg P) for standard curve

Flask	mL std taken	%T*	A_{650}	mg P
Blank	0	100.0	0.000	– 0 –
1				
2				
3				
4				
5				

*Record only if using analog spectrophotometer (Spectronic 20 with meter)

C. Equation of straight line for standard curve

II.   Analysis of fertilizer

A. Mass of fertilizer taken for analysis

mass of sample & weigh boat	
mass of weigh boat	
mass of sample in g	
mass of sample in mg	

B. Analysis of triplicate samples

Volume taken from 100–mL graduated cylinder: _____

Flask	%T*	A_{650}	mg P (from graph)
Analysis 1			
Analysis 2			
Analysis 3			
Average mg P			

*Record only if using analog spectrophotometer (Spectronic 20 with meter)

C. Determination of phosphorus content as %P by mass in the fertilizer

mg P in fertilizer (use average value from above)

$$\text{mg P in sample} = \left(\frac{[\text{mg P (from graph)}]}{[\text{mL sample analyzed}]} \right) \times \left(250 \text{ mL} \times 10 \right)$$

%P in fertilizer

$$\%P = \frac{[\text{mg P in sample}]}{[\text{mg sample}]} \times 100\%$$

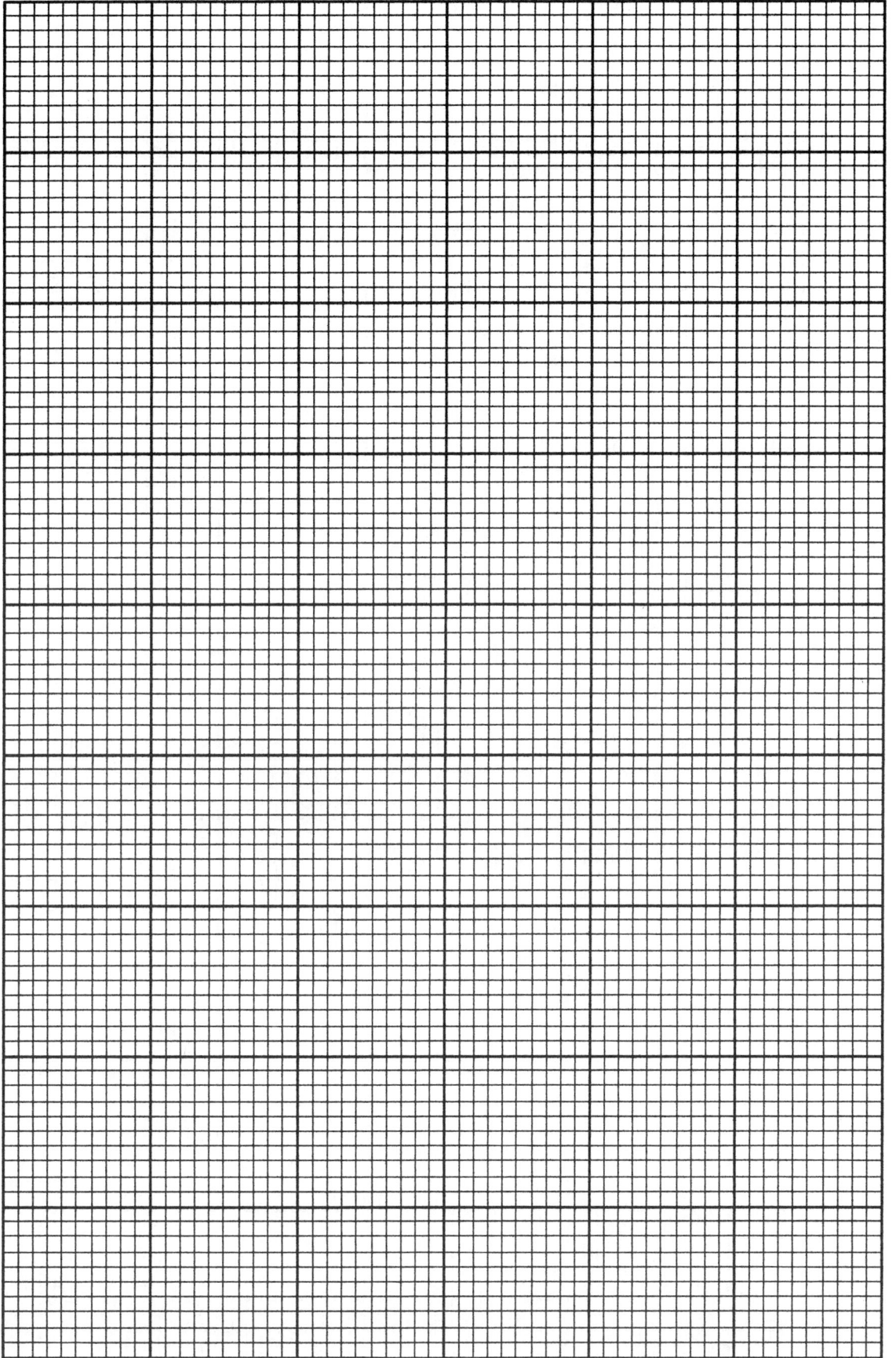

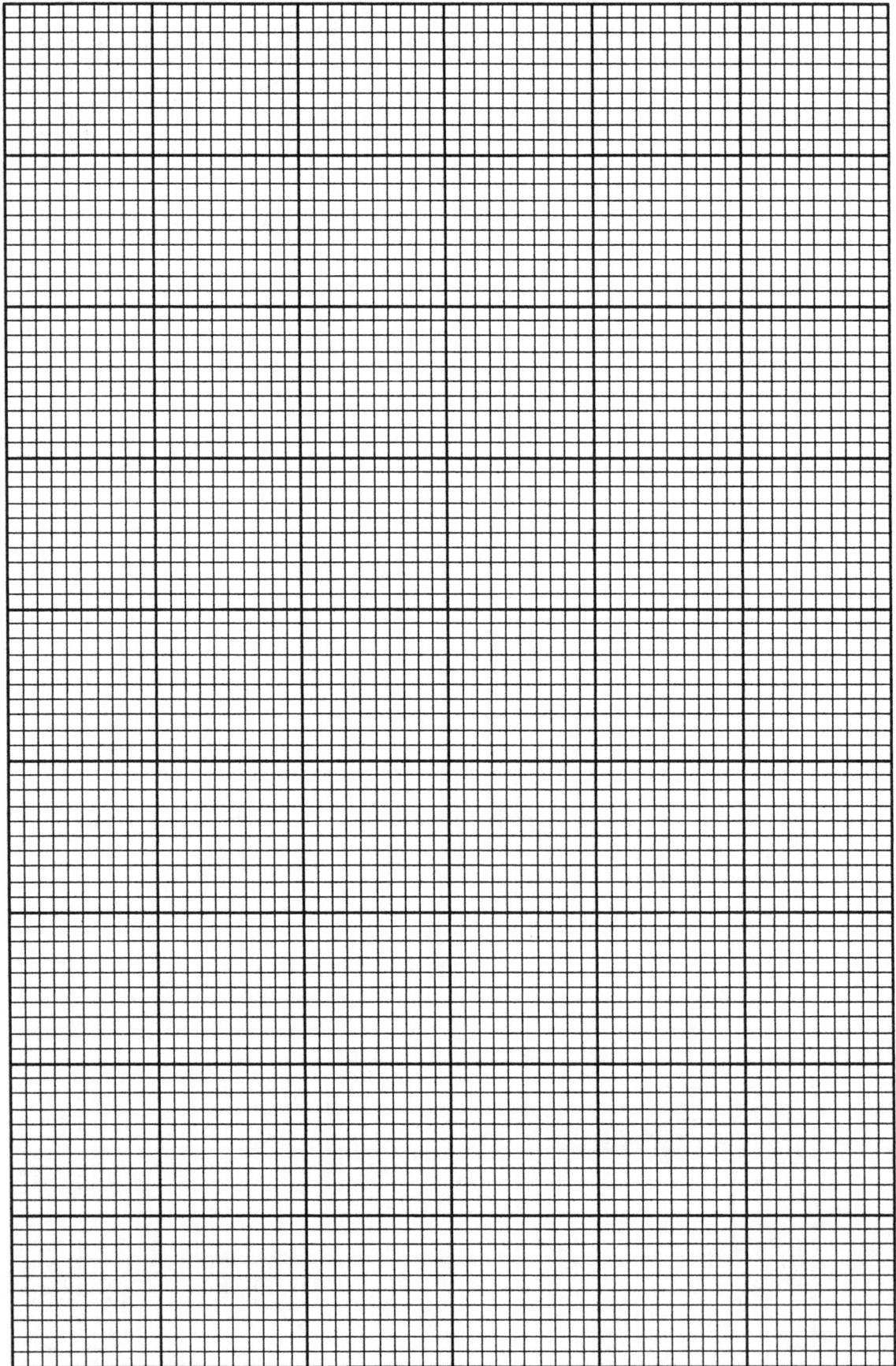

Hard & Soft Water

Purpose

The objectives of this experiment are to examine the properties of hard and soft water and to determine the hardness of a standard water sample and of tap water.

Equipment Needed

five test tubes, test tube rack, 25–mL buret, 10–mL and 100–mL graduated cylinders, two 125–mL Erlenmeyer flasks, droppers or disposable pipets, miscellaneous beakers for solutions and wastes

Reagents Needed

hard and soft water samples, pH 10 buffer, indicator solution (Eriochrome Black T or Calmagite), solutions of soap and detergent

Discussion

Natural waters contain dissolved substances such as gases from the atmosphere and ionic and molecular compounds from the rocks and soil over and through which these waters move. The dissolved substances impart various properties to surface and ground waters, and this experiment explores the chemistry of the characteristic known as *hardness*.

Hardness originally referred to how soaps behaved in water. Soaps mixed with *hard* water formed precipitates, whereas no precipitate appeared when soaps were used with *soft* water. Hard water produces rings of soap scum in bathtubs and causes trouble in getting soaps to lather well. Soft waters leave no bathtub ring and readily form sudsy solutions.

Synthetic detergents do not precipitate in hard water and thus overcome problems caused by soaps. Some synthetic detergents are not biodegradable, however, and hence pass through sewage treatment plants unaltered where they adversely affect aquatic life and the quality of the natural surface waters in streams, rivers and lakes.

Analyses of hard water reveal that these waters contain dissolved cations of calcium, magnesium and occasionally iron. In soft water, monovalent (*i.e.*, 1+) sodium and potassium ions replace the multivalent cations (*i.e.*, 2+ and 3+). Major anions in natural waters are bicarbonate, chloride and sulfate. Hardness depends only on the type of cations present, and to simplify discussions of hard and soft water, you can consider hard water to contain Ca^{2+} and Mg^{2+} (or even just Ca^{2+}) ions and soft water to have Na^+ ions.

Hard water also causes a serious problem in boilers or any circulating hot water system. A rock–hard scale forms within pipes and chambers of these systems, and this scale reduces or even shuts off water flow and dramatically reduces heat exchange. Analyses of the scaly deposits show that they are mostly calcium carbonate. A reaction that represents the formation of boiler scale is

$$Ca^{2+}(aq) + 2 HCO_3^-(aq) \rightarrow CaCO_3(s) + CO_2(g) + H_2O(\ell)$$

Hardness is reported as if all the cations were Ca^{2+} ions and further that all these cations were present in $CaCO_3$. Typical values for natural waters in Georgia range from 1 to 61 parts per million (ppm) $CaCO_3$ which places it in the soft category based on a typical classification shown in the table below. Alabama and South Carolina waters range from 60 to 120 ppm; Florida, 120 to 180 ppm; and some midwestern states, >180 ppm.

Classification	ppm CaCO$_3$	Classification	ppm CaCO$_3$
Soft	< 50	Hard	150 – 300
Moderately hard	50 – 150	Very hard	> 300

Although *parts per million* is actually a mass ratio (such as 1 g per million g), environmental chemists often report concentrations of mg of solute per liter of solution as ppm. (Is this approximation reasonable? What is the ratio of the mass of 1 mg of solute to the mass of 1 L of water?)

The content of Ca^{2+} and Mg^{2+} ions in a sample of water can be determined by titration with a complexing substance called EDTA. The complete name of EDTA is ethylene diamine tetraacetate, and its structure is shown below.

Ethylene diamine tetraacetate

The molecule is flexible and forms a 1:1 complex with multivalent cations such as Ca^{2+} and Mg^{2+} ions. This reaction is represented by the reaction scheme shown on the next page; the complex is also depicted as a line structure. In this structure, carbon and hydrogen atoms are omitted and implied to be at the line intersections. An identical reaction occurs with Mg^{2+} ions and EDTA. The unshared pairs of electrons on the two nitrogen atoms and four of the oxygen atoms bind to multivalent cations and hold them as if in a cage. Monovalent cations, however, do not form strong enough bonds to be trapped.

Indicators, such as Eriochrome Black T and Calmagite, are available to detect when the reaction with multivalent cations is complete. These indicators themselves are complexing agents which bind Ca^{2+} and Mg^{2+} ions more weakly than EDTA. These indicators exhibit one color when a cation is bound and change to a different color when the cation is removed. At the first of a titration the indicator exists in the bound complex form. When EDTA has formed a complex with all the free Ca^{2+} and Mg^{2+} ions in solution, the next drop of EDTA removes the cations from the indicator which then changes color signifying the end of the reaction.

In this experiment, you will first examine the properties of hard and soft waters when mixed with soap and synthetic detergents. Then you will determine the hardness of a prepared water sample and tap water by EDTA titrations.

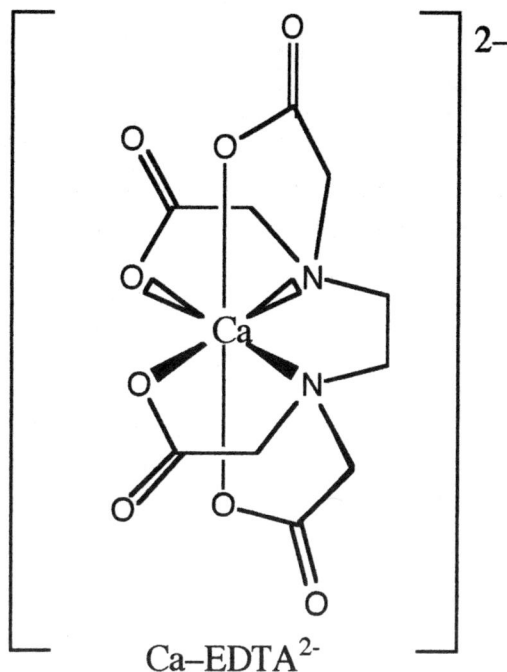

$$Ca^{2+} + EDTA^{4-} \longrightarrow Ca\text{--}EDTA^{2-}$$

Ca–EDTA^{2-}

Procedure

I. Soaps and detergents in hard and soft water

Add 2 to 3 mL of soap solution to two test tubes and 2 to 3 mL of detergent to two other test tubes. Add a few drops of *soft* water to a tube containing soap and a few drops to a tube containing detergent. Mix the contents well. Repeat the process adding a few drops of *hard water* to soap and detergent solutions.

Set the tubes aside in the test tube rack, and observe any changes that may occur during the rest of the laboratory. Record your observations on your laboratory report.

II. Analysis of Ca^{2+} and Mg^{2+} ions in hard water

Fill a 25–mL buret with 0.0100 M EDTA solution. Make certain that there is no air in the tip of the buret.

Carefully measure 50.0 mL of the water sample in a graduated cylinder and transfer it to a 125–mL flask. Add 1 mL of the pH 10 buffer and 2–4 drops of the indicator.

Titrate with the standard 0.0100 M EDTA solution until the color changes from red to **pure** blue. Record the initial and final buret readings on the data sheet. Repeat the titration two more times so that you will have three analyses to average.

Calculate the hardness of the water sample assuming that Ca^{2+} ions cause all the hardness. Use the principles of solution stoichiometry that you have learned in class and in other laboratories involving volumetric analyses. Report the hardness in terms of the molarity of Ca^{2+} ions in the water as the average of your triplicate analyses.

Environmental chemists and engineers convert the molarity values of the Ca^{2+} ions to **mg $CaCO_3$ per L of sample**. This number is the same as **ppm $CaCO_3$**. Use the molar mass of $CaCO_3$ and the appropriate conversion factors for this calculation. Consult with your laboratory partner to verify that you are proceeding correctly.

III. Hardness of tap water

To determine the hardness of tap water, you may have to run trial analyses to obtain conditions that will produce optimal experimental numbers. Two parameters are easily adjustable—volume of tap water taken for analysis and concentration of the EDTA solution.

The volume of the sample measured (in a graduated cylinder in this experiment) must be large enough so that slight errors in reading the measuring device or in completely transferring the sample are insignificant. For example, you want to take at least a 25–mL sample from a 100–mL graduated cylinder and preferably 50 to 100 mL.

Also for accuracy you need to deliver 20 to 24 mL from the 25–mL buret to reach the end point in a titration. (Because time does not permit you to prepare EDTA solutions of different concentrations, you may extend this range from 10 to 24 mL.) An error of 0.1 mL in reading the initial and final volumes on a buret is a much smaller percent error when 24 mL have been delivered than when only 5 mL have been used.

First, repeat the titration using the *same* volumes used in the titrations of the standard hard water sample. If you find that a 50–mL sample of tap water uses 10 to 25 mL of the EDTA solution, you do not have to vary any parameters (*i.e.*, you got lucky).

Repeat the procedure twice more on tap water so that you have triplicate analyses.

If you find that a 50–mL sample of tap water requires *more* than 25 mL of EDTA, reduce the volume of tap water taken for analyses.

If you find that a 50–mL sample of tap water consumes *less* than 10 mL of the EDTA solution, then you must dilute the EDTA solution. Based on your trial titration, calculate how much EDTA solution a 50–mL sample of tap water would require if the EDTA were 0.005 M or 0.0001 M. These solutions are prepared by diluting 50 mL to 100 mL or 10 mL to 100 mL with deionized water. Once you have determined a suitable concentration of EDTA to use in the titration, run triplicate analyses on a 50–mL sample of tap water.

Complete the calculations of your triplicate analyses and report your results in terms of **mg $CaCO_3$ per L of tap water** or **ppm $CaCO_3$**.

IV. Cleanup

All solutions may be rinsed down the drain with tap water. Rinse the burets with deionized water, open the stopcocks and leave them inverted on their stands to drain.

Laboratory Report
Experiment 11
Hard & Soft Water

Name: _____

Date & Time: _____

Lab Partner: _____

I. Soaps and detergents in hard and soft water

    A. Observations

        soft water and soap

        soft water and detergent

        hard water and soap

        hard water and detergent

    B. Conclusions

II. Analysis of Ca^{2+} and Mg^{2+} ions in hard water

Molarity of EDTA solution: _____

Measurements	Titration 1	Titration 2	Titration 3
Volume of water sample			
Final buret reading			
Initial buret reading			
Volume used (mL)			

Calculations	Titration 1	Titration 2	Titration 3
Moles of EDTA used			
Mol Ca^{2+} reacted			
Molarity of Ca^{2+} ions			
Average Molarity			
Hardness as mg $CaCO_3$ / L			

III. Hardness of tap water

A. Trial titration(s)

Trials	1	2	3	4
Volume tap water used				
M of EDTA used				
mL EDTA used				

B. Tap water titrations

Molarity of EDTA used for tap H_2O titrations: _____

Measurements & Calculations	Titration 1	Titration 2	Titration 3
Volume tap water taken			
Volume EDTA used			
Molarity of Ca^{2+} ions			
Average Molarity			
Hardness as mg $CaCO_3$ / L			

Corrosion

Purpose

This experiment explores how iron corrodes. The scientific objectives are to examine the effect of acidity on the oxidation of iron and to examine ways to prevent or accelerate the corrosion of iron.

Equipment Needed

test tubes, test tube rack, petri dishes, stirring rod, beakers (250–mL, 400–mL, 600–mL), tweezers, 24–well spot plate, two Beral pipets, miscellaneous flasks for solutions and wastes, hot plate, burner with ring stand, ring and wire gauze, tongs, Parafilm

Reagents Needed

iron nails (8 to 10), deionized water, oil, clear nail polish, samples of Mg, Cu and Zn, universal indicator, small piece of concrete, 2 M H_2SO_4 and 0.1 M solutions of NaOH, Na_2CO_3, KNO_3, $AlCl_3$ and HNO_3

Discussion

When metals are exposed to the environment, they deteriorate through a process known as *corrosion*. Oxidation–reduction reactions cause metals to corrode, and the resulting products diminish the structural strength of the metal.

Most metals form metal oxides when they corrode. Corrosion of iron produces the familiar reddish–brown iron(III) oxide, and iron is said to *rust*. Aluminum corrodes by readily forming a thin greyish–white film of aluminum oxide on its surface when exposed to air. Copper and silver also corrode when left out in the environment, but the resulting compounds are not oxides but green–colored copper(II) carbonate and dark brown–black silver sulfide. These latter two metals are said to *tarnish*.

Preventing corrosion and replacing iron that has corroded occupy major portions of our economy. The repair and replacement of rusting iron, for example, consumes approximately 25% of the annual production of steel in this country. Much of the output of the paint and coatings industry also goes to prevention of corrosion.

In corrosion processes, the metal serves as the source of electrons and oxygen from the atmosphere is the recipient or ultimate receptor of these electrons lost by the metal atoms. The oxidation and reduction reactions are physically separated, and electrons flow through the metal from the atoms losing them to the oxygen molecules accepting them. These separate half reactions for the corrosion of iron are depicted below.

In these reactions, the ionic products initially formed combine to produce a precipitate of iron(II) hydroxide.

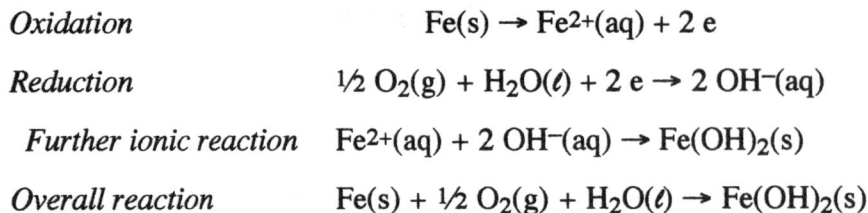

Oxidation	$Fe(s) \rightarrow Fe^{2+}(aq) + 2\,e$
Reduction	$\frac{1}{2}\,O_2(g) + H_2O(\ell) + 2\,e \rightarrow 2\,OH^-(aq)$
Further ionic reaction	$Fe^{2+}(aq) + 2\,OH^-(aq) \rightarrow Fe(OH)_2(s)$
Overall reaction	$Fe(s) + \frac{1}{2}\,O_2(g) + H_2O(\ell) \rightarrow Fe(OH)_2(s)$

Common experience tells us that iron rusts more readily when it is wet or in water than when it is kept dry. The reduction half of the reaction reveals that water and molecular oxygen are needed, and water in contact with the atmosphere contains sufficient levels of dissolved oxygen to facilitate this reaction. Also acids react with iron(II) hydroxide and shift the overall reaction in favor of products, and iron rusts faster under acidic conditions than in neutral or basic solutions as anyone who has spilled battery acid on their car can tell you.

The iron(II) oxide that forms may undergo further oxidation if an ample supply of oxygen and water are available (as they usually are) to produce hydrated iron(III) oxide as shown below. Iron(III) oxide is reddish–brown.

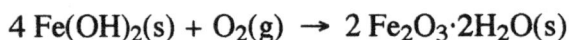

$$4\,Fe(OH)_2(s) + O_2(g) \rightarrow 2\,Fe_2O_3 \cdot 2H_2O(s)$$

Variable amounts of water associate with iron(III) oxide, and *rust* is usually represented as $Fe_2O_3 \cdot (H_2O)_x$ where x is 2 to 4.

In this experiment, you will investigate conditions that promote and inhibit corrosion of iron nails. You will also have the opportunity to design your own experiments to study the corrosion of iron. There are two parts to the common experiments.

In the first part, you will observe the corrosion of an iron nail in a jelly–like medium called agar. Dissolved in the agar is phenolphthalein and potassium ferricyanide which has the formula $K_3Fe(CN)_6$. Phenolphthalein is an acid–base indicator that is pink in basic solution and colorless in acidic solution. Potassium ferricyanide furnishes $Fe(CN)_6^{3-}$ ions that combine with Fe^{2+} ions from the rusting iron to form a blue–colored compound known as Prussian blue $Fe_3[Fe(CN)_6]_2$. If you examine the half reactions for the corrosion of iron and then consider the colored products with phenolphthalein and ferricyanide ions, you can identify where the half reactions occur on the iron nail. Agar prevents the rapid mixing of the oxidation–reduction products so that the half reactions can be localized.

In the second part of the experiment, you will observe changes that occur when iron nails are placed in acidic, basic and neutral solutions. You will estimate acidity on aliquots of the solutions with a universal indicator. The color of this indicator corresponds to a range in the *pH* of a solution, and pH is simply a measure of acidity. Pure water has a pH of 7 and is considered *neutral*. *Acidic* solutions have pH values less than 7, whereas *basic* solutions have pH values greater than 7. You can read more about pH in your textbook.

Lastly you will design and test two treatments of iron nails under conditions of your own choosing. Suggestions will be made, but you are encouraged to develop your own ideas of how to investigate corrosion of iron.

Procedure

I. Preparing clean iron nails

Bring about 200 mL of deionized water to boiling in a 400–mL beaker. Obtain 10 iron nails and lay them flat in a 250–mL beaker. Cover the nails with about 25 mL of 2 M H_2SO_4. After the nails have been in the acid for at least 5 minutes, decant the acid into a waste container. Rinse the nails with copious amounts of tap water and finally with two or three rinses of deionized water. Then place the nails in the boiling deionized water. Remove the nails with tweezers and dry them with a clean towel **as you need them.** (The hot deionized water will prevent the cleaned nails from oxidation. Can you suggest a reason why this protection occurs?)

II. Observing the half reactions of iron nails rusting

A. Preparation of agar media

A large beaker of agar will usually be prepared for the laboratory. Be careful! The agar is hot and sticky. With your instructor's permission, however, you may prepare a small quantity of agar as follows.

Heat 100 mL of deionized water to boiling in a 250–mL beaker. Remove the burner, add 1 g of powdered agar to the boiling water, and stir the solution vigorously with a stirring rod to disperse agar in the solution. Add 10 drops of 0.1 M $K_3Fe(CN)_6$ and 10 drops of 0.1% phenolphthalein indicator to the dissolved agar and thoroughly mix the contents of the solution. Allow the agar to cool slightly before using. If the agar forms a gel before you need to pour it, you can liquify it again by gentle heating.

B. Preparation of petri dishes to observe iron corrosion

Place an acid–cleaned iron nail in a petri dish (100 x 15 mm). Pour about 40–50 mL of agar solution into the dish making certain that the nail is completely covered. Place the lid on the petri dish and label it. Observe the nail during the course of the laboratory period and set the dish aside until next week when you can complete your observations.

III. Effect of acidity on corrosion of iron

Place a clean iron nail in each of seven test tubes and cover the nails with a different solution. The 0.1 M solutions to use are NaOH, Na_2CO_3, KNO_3, $AlCl_3$, and HNO_3. In another tube, use deionized water only. In the last tube, place a small piece of concrete into deionized water. Immerse the nails completely in the solutions.

To determine the acidity of each tube, add about 10–20 drops of each solution to separate wells of a 24–well plate. Place a drop of universal indicator in each well and record the color of the resulting solution. Compare the color with the pH color values for the indicator, and record your analysis of the acidity of each solution.

Observe changes in the nails during the laboratory period and then set the tubes aside to make your final observations at next week's laboratory.

IV. Investigation of ways to prevent or accelerate corrosion of iron

Design **at least two** experiments to determine the effects of various treatments to prevent or accelerate corrosion of the iron nails. Possible treatments that you may want to consider are listed below. You may use agar medium in petri dishes or solutions in test tubes for the test environment.

1. Protective coatings—Coat a treated nail with oil and/or clear fingernail polish (paint substitute). A variation would be to coat only part of the nail.

2. Active metal protection—Tightly wrap Cu wire or Mg ribbon around one end of a nail. A variation would be to cover part of the nail and the attached metal with fingernail polish.

3. Stress points in metal—Sharply bend or crimp a nail.

4. Annealing—Annealing is the process of slow cooling of a metal that prevents stress points from occurring. Heat the end of a nail in the hottest part of the flame of a Bunsen burner. Plunge the nail into a beaker of water, and then heat it again in the flame. **Slowly** cool the iron before using it in an experiment.

5. Oxygen–free water—Boil a small amount of water and immediately fill a test tube completely with it. Quickly place a cleaned nail into the test tube and seal the tube with a rubber stopper so that no air is trapped in the tube.

Briefly outline your procedure on the report sheet, and record your **results** and **conclusions**.

V. Cleanup and waste disposal

Remove all nails, rinse them with tap water and leave them on towels to try. Place solidified agar in the trash. Dispose of the $AlCl_3$, HNO_3 and H_2SO_4 solutions in the indicated waste beakers. All other solutions can be rinsed down the drain with tap water.

Laboratory Report
Experiment 12
Corrosion

Name: _____

Date & Time: _____

Lab Partner: _____

I.   Observation of oxidation–reduction half reactions of iron nails rusting

A. Appearance of agar with clean nail at start of experiment

B. Observed changes in the agar during the laboratory

C. Places along nail where the following reactions occur (write the balanced half reactions)

1. Oxidation

Balanced half reaction: _____

Location on nail:

2. Reduction

Balanced half reaction: _____

Location on nail:

IV. Calculation of R_f values and identification of the "evidence" and commercial samples

For each spot, measure the distance from the origin to the center of the spot. Also measure the distance from the origin line to the solvent front. Calculate R_f values by dividing the distance each compound migrates by the distance the solvent moves.

Use the R_f values of known standard compounds to identify the components in the "evidence" and in the commercial samples. Compounds often do not produce a compact spot on the chromatogram but form an elongated smear. Also the solvent may not move up the plate evenly. What causes these conditions is unclear and hard to predict and control. Hence, you may have to repeat the chromatography until you obtain chromatograms that you can interpret.

Apply scientific reasoning in making your conclusions. Do the R_f values of all the components in the "evidence" correspond to R_f values of known compounds? How do the R_f values of the components in the commercial preparations compare to the R_f values of the pure compounds? Record what you can conclude from analysis of your chromatograms as well as what you cannot determine. You must prove or justify your conclusions based on your chromatograms.

II. Effect of acidity on corrosion of iron

A. Data table

Solution	Indicator Color	pH Range	Acidity*	Observations
NaOH				
Na$_2$CO$_3$				
KNO$_3$				
AlCl$_3$				
HNO$_3$				
H$_2$O				
Concrete & H$_2$O				

*Classify solution as very basic, moderately basic, neutral, moderately acidic, or very acidic.

B. Analysis and Conclusions

III. **Investigation of two ways to prevent or accelerate corrosion of iron**

A. First investigation

1. Experimental design *(What did you do?)*

2. Observations *(What happened?)*

3. Conclusions *(What does what you did mean?)*

B. Second investigation

   1. Experimental design *(What did you do?)*

   2. Observations *(What happened?)*

   3. Conclusions *(What does what you did mean?)*

Drug Bust!
Identifying an Unknown Powder
by TLC

Purpose

The purpose of this experiment is to determine whether an unknown powder contains common headache medicines as claimed by an accused drug dealer. [OK, so it's not real evidence from a real crime, but let's pretend a little here. The methods and scientific reasoning are analogous to what is done in analytical laboratories and crime laboratories.]

Equipment Needed

silica gel thin–layer plates containing fluorescent indicator, capillary tubes, test tubes and test tube rack, Beral pipets, chromatography jar (600–mL beaker with Saran Wrap), mortar and pestle, ultraviolet lamp

Reagents Needed

acetylsalicylic acid, acetaminophen, ibuprofen, caffeine, unknown powder (the "evidence"), ethyl acetate, commercial tablets such as Anacin, Bayer Aspirin, Excedrin, Advil and Tylenol

Discussion

[With apologies to Batman!] You have just arrived at your job in Gotham City's crime laboratory when the DA bursts into your office with a vial containing a white powder. He plops the vial on your desk and says "Two caped men wearing funny suits caught a guy, who looks like a penguin, selling this stuff on the street corner in an area known for illicit aspirin sales. The 'perp' claims it contains no aspirin but is just powdered headache medicine which he prepares and sells as a community service to passersby who are too busy to stop at a pharmacy. What do you think? You scientists have got to help us on this one." You gulp down some day–old coffee and say, "Sure, Chief. I'll analyze the powder and get back with you later this afternoon." The DA leaves, and your thoughts turn from dames and dudes and Saturday nights to chemistry, the laboratory and thin–layer chromatography.

In this laboratory, you will analyze the contents of a powder and use the technique of thin–layer chromatography to identify its components. Thin–layer chromatography, also referred to as TLC, is a powerful method for the separation of mixtures and identification of their components.

TLC consists of a thin film of silica gel coated on a glass plate or plastic strip, and silica gel traps water, a polar solvent, within its three–dimensional network. Samples are applied to the thin–layer plate in a small dot; only 1 to 10 μL of sample are needed. When the edge of the plate is immersed in a solvent, the solvent slowly creeps up the plate and over the applied sample. The solvent is selected to be less polar—hence, *more nonpolar*—than water.

The solvent and the bound silica gel with its trapped water create two phases—one stationary and one mobile. The mobile solvent and the stationary silica gel are selected so that the compounds in the sample have different affinities for the two phases. As the solvent sweeps up the plate, compounds that are more soluble in it than in the silica gel phase move up the plate. Compounds that prefer the stationary silica gel with its trapped water lag behind. The overall result is that when the solvent gets near the top of the plate, the compounds in the mixture are strung out according to their solubilities in the two phases.

At this point, the plate is removed from the solvent, and the solvent front is marked with a pencil. The plate is placed in a ventilated cabinet where the solvent evaporates. Then the locations of the compounds are identified using some of their physical or chemical properties. In this experiment, the silica gel is impregnated with a fluorescent compound. When ultraviolet light is shined on the plate, this compound fluoresces and gives the plate a light purple color. Many organic compounds quench or greatly reduce the fluorescence. Hence in ultraviolet light, the separated compounds appear on the plate as dark spots surrounded by purple. The figure below depicts three stages in the chromatography of compounds A and B.

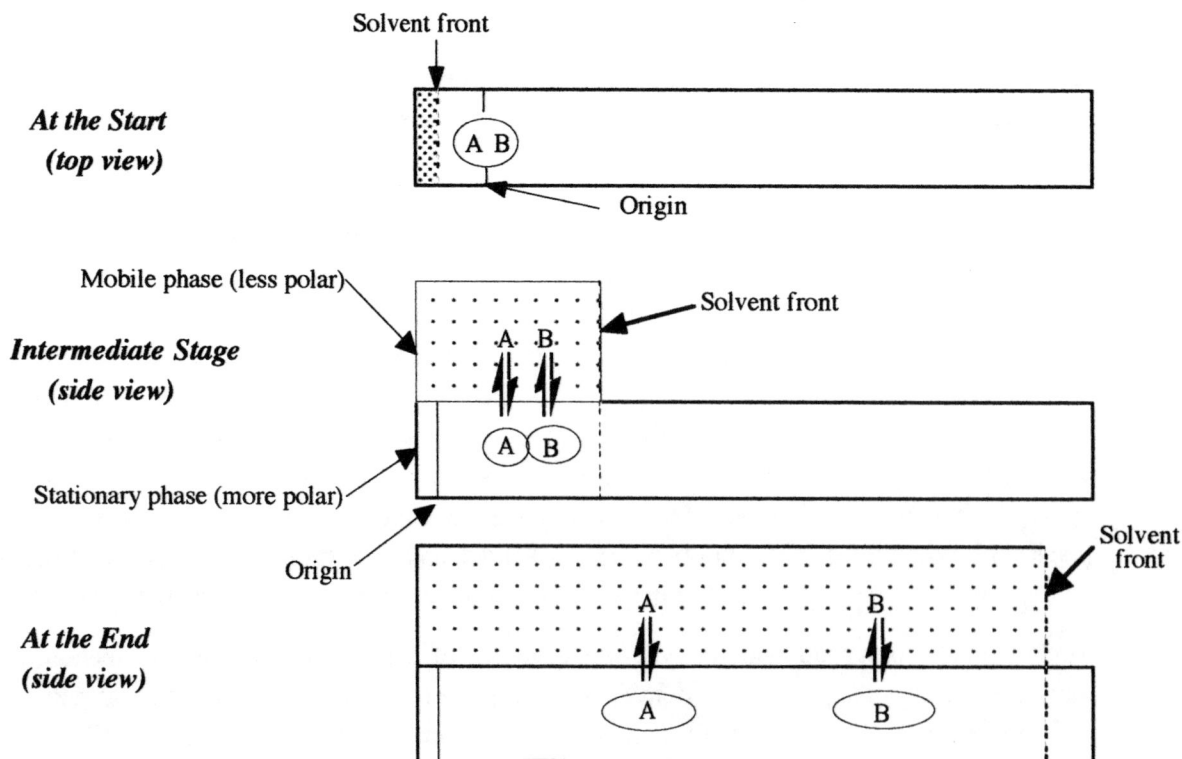

Compound B migrates farther from the origin that compound A and thus is more soluble in the nonpolar mobile phase than in the polar water trapped in the silica gel.

The ratio of how far a particular compound migrates compared to the distance the solvent moves is called the **R_f** factor. For a particular solvent and a particular stationary phase, each compound will have a characteristic R_f factor. Compounds are identified by comparing their chromatographic behavior in a particular system to that of known compounds in the same system. The diagram below shows how R_f values are calculated.

Top view of TLC plate with measurements for R_f Calculations

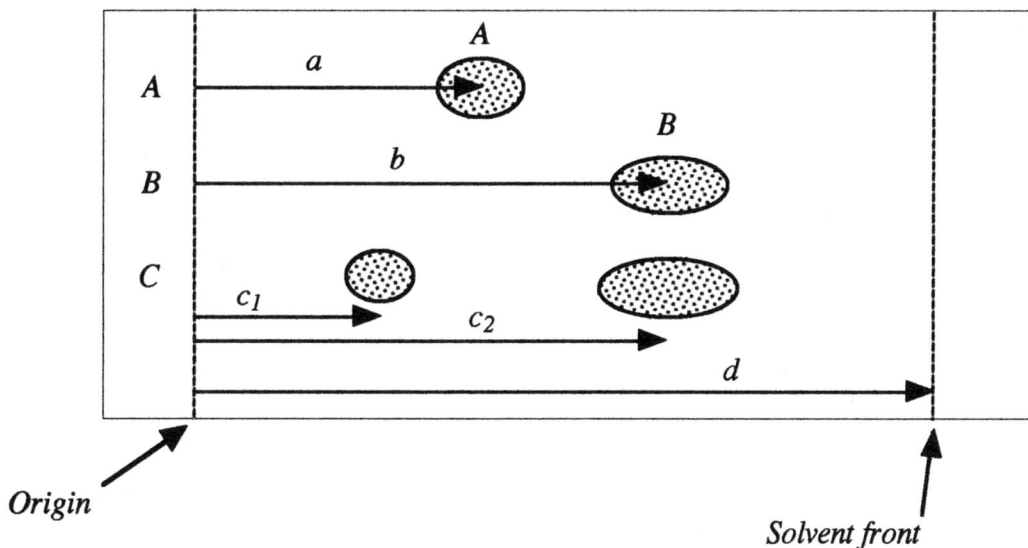

The following ratios represent the R_f values for samples spotted at A and B.

$$R_f(A) = \frac{a}{d} \qquad\qquad R_f(B) = \frac{b}{d}$$

The sample spotted at C appears to consist of two compounds. Each compound has an R_f value calculated as

$$R_f(C_1) = \frac{c_1}{d} \qquad\qquad R_f(C_2) = \frac{c_2}{d}$$

Based on the apparent identical values of $R_f(B)$ and $R_f(C_2)$, you would identify C_2 as compound B. Compound C_1 is unknown from its chromatographic behavior. Occasionally, compounds do not form compact spots but long smears. In those cases, you estimate the center of the smear in taking the measurements.

In this experiment, you will compare the TLC behavior of the unknown powder representing the "evidence" in the crime laboratory scenario to the behavior of four purified compounds (called the *standards*) and to the behavior of common nonprescription formulations of these compounds (such as Aspirin, Anacin, Tylenol, Advil and Excedrin). Remember that the purpose of the analysis is to identify what compound or compounds are present in the unknown powder, and if possible, suggest what common over–the–counter drug it might be. You will base the identification of the components in the unknown powder on R_f values of standard pure compounds and the components of commercial tablets.

The names and structures of the possible compounds are shown below. You can read more about chemical structures in the section on organic chemistry in your textbook or in texts on organic chemistry. The nonprescription tablets that you will analyze contain one or more of these compounds which act as analgesics, anti–inflammatory agents and stimulants. Analgesics are painkillers and include aspirin and acetaminophen. Aspirin and ibuprofen exhibit anti–inflammatory properties. Caffeine is a stimulant.

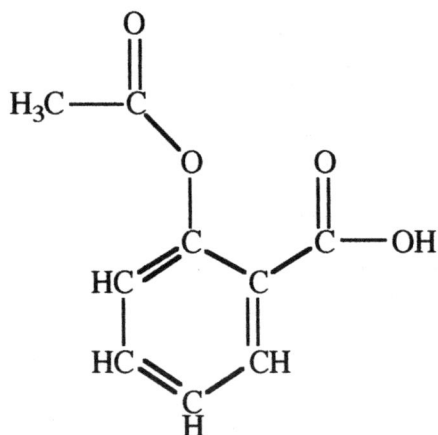

Acetylsalicylic acid (aspirin)

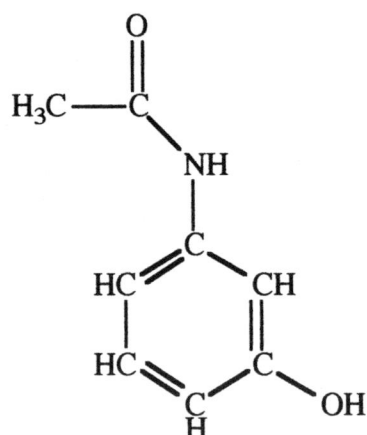

Acetaminophen

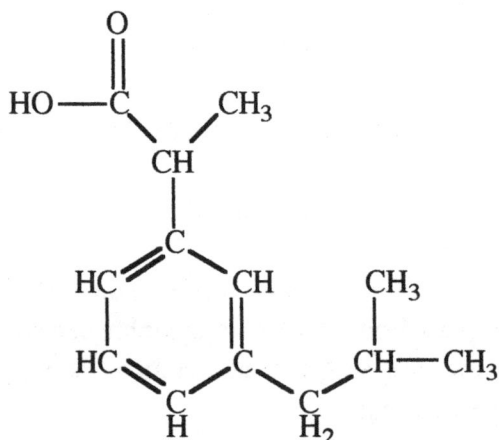

Ibuprofen

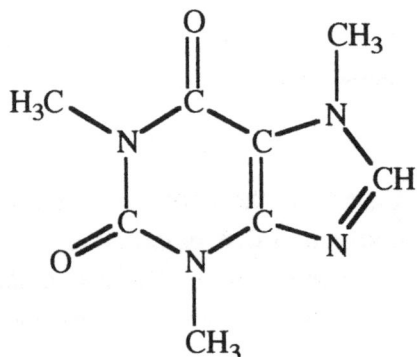

Caffeine

Procedure

I. Preparation of samples

Work with neighboring groups at your laboratory bench and prepare samples of each commercial product as follows. In a clean **and** dry mortar, grind one tablet to a fine powder and transfer the powder to a clean test tube. You will have enough powder to share with 2 or 3 groups. Collect samples of each commercial tablet to be analyzed in labeled test tubes.

Also place in labeled tubes small amounts (about the size of a 2–mm pellet) of each of the purified standard compounds—acetylsalicylic acid, acetaminophen, ibuprofen and caffeine. In another tube add a small amount of the "evidence." Record the number on the "evidence" on your laboratory report.

To each tube, add 1.0 mL of methanol using a Beral pipet. Thoroughly mix the contents of each tube, and allow any insoluble material to settle.

II. Application of samples to TLC plates

Obtain two thin–layer plates and handle them by the edges only. Across one end of each plate, lightly draw a line with a pencil about 1 cm from the end. Place hatch marks at 8–mm intervals along this line. You can get 5 samples on a 5–cm wide plate using these measurements.

On the first plate, chromatograph the evidence and the four standard compounds. Take a capillary tube and touch it to the liquid in the sample tube being analyzed. **Gently** touch the end of the capillary tube to the plate at one of the hatch marks. Let some of the liquid flow from the capillary to the plate. The size of the spot should be 2–3 mm in diameter. Allow the spot to dry and then reapply the sample to the same spot. Repeat this process with other samples using a clean capillary tube for each sample. Record the position of each sample or lightly write on the plate with a pencil.

On the second plate, chromatograph the evidence and samples of four commercial tablets. Mark the TLC plate and apply the samples as described above.

III. Chromatography of the samples

Add a few mL of ethyl acetate (or solvent from the reagent bottle labeled *Mobile Phase*) to a 600–mL beaker or chromatography jar. The height of the liquid should be about 0.5 cm.

Place the plate into the 600–mL beaker so that the sample line is **not** submerged in the ethyl acetate solvent. Cover the beaker or jar with Saran Wrap. With care, you can run two plates simultaneously in the same beaker or you can run one after the other. You will have sufficient time to repeat the analyses several times if necessary.

When the liquid is about 80% up the plate, remove the plate, quickly mark the solvent front with a pencil, and let the plate dry in the hood. Do not forget to keep the chromatography jar covered whenever possible.

Laboratory Report
Experiment 13
Drug Bust! Identifying an
    Unknown Powder by TLC

Name: _____

Date & Time: _____

Lab Partner: _____

I. **Sketch of thin–layer plates after chromatography**   *Evidence No.* _____

  A. Evidence and standard compounds   B. Evidence and commercial tablets

II. **R$_f$ values of the "evidence," purified compounds and commercial tablets**

Sample	R$_f$ Value(s)
Evidence	
acetylsalicylic acid	
acetaminophen	
ibuprofen	
caffeine	

Sample	R$_f$ Value(s)
Evidence	

III. Analysis and Conclusions

Write a brief summary of your analysis of the TLC plates. Identify as many compounds in the evidence as you can. Also report on what compounds are in the commercial tablets **based on your analyses**. Can you conclude whether or not the accused was selling a commercial product? Can you conclude whether or not the "perp" was selling aspirin or an aspirin–free product?

IV. Your *expert* testimony says: The person accused of selling a product that contained aspirin was

Guilty: _____   **Not Guilty:** _____

Hot & Cold Packs

Purpose

The objectives of this experiment are to determine the heats of solution of several ionic compounds and then use this information to design hot and cold packs for specified temperature changes.

Equipment Needed

Standard procedure: coffee–cup calorimeter (two styrofoam cups nestled inside each other) and lid, thermometer, glass stirring rod, 100–mL graduated cylinder

Computer–interface procedure: coffee–cup calorimeter with lid, magnetic stirrer with teflon–coated stir bar, computer interface with temperature probe, computer

Reagents Needed

magnesium sulfate, potassium nitrate, ammonium nitrate

Discussion

The heat of solution (also called *enthalpy of solution*) is the heat that is released or absorbed when an ionic compound (frequently referred to as a *salt*) dissolves in water. The symbol for this heat is ΔH_{soln} , and its units are $kJ \cdot mol^{-1}$. An example of this type of process for the dissolution of potassium chloride in water is represented symbolically as

$$KCl(s) \xrightarrow{\text{H}_2\text{O}} K^+(aq) + Cl^-(aq) \qquad \Delta H_{soln} = +17.5 \text{ kJ} \cdot mol^{-1}$$

Heat exchanges for these processes are measured by constant–pressure calorimetry. *Experiment 8* discusses the necessary measurements and equations used to calculate this enthalpy change, and you may need to consult that experiment to refresh your memory.

The value of ΔH_{soln} enables you to estimate how much of the salt would be needed to produce a desired temperature change. This information can thus be used in making hot packs (if the process is exothermic) or cold packs (if the process is endothermic).

For example, assume that you want to make a cold pack that has a total mass of 400 g and will lower the temperature of the solution that forms to 10°C. Typical room temperatures are 23°C, and thus the change in temperature ΔT is –13°C. In this type of calculation, you are only interested in the absolute values of ΔT and not in the direction of heat transfer. Thus you use 13°C in the calculations. Further you can assume that the specific heat of the solution is that of water $4.184 \text{ J} \cdot g^{-1} \cdot °C^{-1}$.

The heat that must be absorbed is calculated from the equation below assuming that the heat capacity of the calorimeter is negligible.

$$q_{soln} = m_{soln} \cdot C_{soln} \cdot \Delta T$$

Substituting these needed quantities into this equation gives the following value of heat that must be absorbed to produce the desired temperature change.

$$q_{soln} = (400 \text{ g})(4.184 \text{ J} \cdot \text{g}^{-1} \cdot {}^{\circ}\text{C}^{-1})(13{}^{\circ}\text{C})$$

$$q_{soln} = 2.18 \times 10^4 \text{ J}$$

The endothermic process for the dissolution of KCl shown in the reaction scheme above absorbs 17.5 kJ of heat for every mole of KCl that dissolves. The molar mass of KCl is 74.55 g, and thus the process absorbs 0.235 kJ or 235 J per gram of KCl.

Dividing the calculated q_{soln} by the heat absorbed per gram of dissolving KCl then indicates how much KCl is needed. Check the mathematics and units of this calculation to see that they make sense.

$$\text{g KCl needed} = \frac{q_{soln}}{235 \text{ J} \cdot \text{g}^{-1} \text{ KCl}}$$

$$\text{g KCl needed} = \frac{2.18 \times 10^4 \text{ J}}{235 \text{ J} \cdot \text{g}^{-1} \text{ KCl}}$$

$$\text{g KCl needed} = 92.7 \text{ g} \ or \ \text{rounded to } 93 \text{ g KCl}$$

The total mass of the cold pack is to be 400 g, and thus when 93 g of KCl is mixed with 307 g of H_2O, you will have a cold pack that reaches 10°C when the water and salt are mixed.

The table below lists literature values for the heats of solution of ionic compounds that you will investigate in this experiment. You will compare your results to these values and then use your values to design 400–g hot and cold packs that reach 60°C and 10°C, respectively.

Ionic Compound	Formula	ΔH_{soln} (in kJ·mol⁻¹)
Magnesium sulfate	$MgSO_4$	–84.9
Potassium nitrate	KNO_3	+35.6
Ammonium nitrate	NH_4NO_3	+26.4

Procedure

I. Standard procedure for calorimetry

Weigh a dry calorimeter (without its cover) to the nearest 0.001 g. Add 100 mL of deionized H_2O from a graduated cylinder, and reweigh the calorimeter with the water. This second mass can only be recorded to the nearest 0.01 g.

Weigh out 5.00 ± 0.01 g of ionic compound under investigation in a plastic weigh boat.

Assemble the calorimeter with its lid, thermometer and stirring rod as shown in the figure on page 56 of this manual.

Record the temperature of the water to the nearest 0.1°C. Remove the cover of the calorimeter and add the ionic compound. Then quickly replace the cover. Observe the temperature and gently stir the solution with the glass rod until the temperature stops changing. Record the constant temperature to the nearest 0.1°C.

After you have completed collecting data, wash the contents of the calorimeter down the sink. None of the compounds under investigation are harmful to the environment. Dry the calorimeter with a paper towel.

Perform this procedure with one salt that dissolves by an exothermic process and one by an endothermic process. If time permits, run duplicate analyses of each salt investigated. Complete the calculations described in part III below as discussed in the introduction to this laboratory.

II. Calorimetry using computer–interface systems

Review the discussion in *Appendix A* of this manual on the use of computer–interface systems in chemical experimentation and the section in *Experiment 8* that describes use of these systems in calorimetry.

Weigh a dry calorimeter (without its cover) to the nearest 0.001 g. Add 100 mL of deionized H_2O from a graduated cylinder, and reweigh the calorimeter with the water. This second mass can only be recorded to the nearest 0.01 g.

Weigh out 5.00 ± 0.01 g of ionic compound under investigation in a plastic weigh boat.

Assemble the calorimeter with its lid, as shown on page 56 but replace the thermometer with the temperature probe attached to the computer–interface and the stirring rod with a teflon–coated stir bar. Arrange the calorimeter on a magnetic stirrer and secure it with an iron ring attached to a ring stand to prevent the calorimeter from accidentally tipping over.

Calibrate the probe using a beaker of ice water containing a thermometer. Follow the instructions on the computer screen. Repeat the process with a beaker of hot tap water for the second calibration point. When the calibration is complete, replace the temperature probe in the calorimeter and adjust the magnetic stirrer so that the solution is stirring gently.

Design the experimental procedure for data collection by selecting a previously written program (or designing one yourself) as indicated by your instructor.

Acquire data for about a minute prior to the addition of the previously weighed salt sample. Add the salt and continue data collection until the temperature has become constant for about 2 minutes. Save your data in a file as instructed.

Analyze your data from the saved file by graphing temperature on the *y*–axis versus time on the *x*–axis. Use the graphical information to determine *initial* and *final* temperatures and record these values on your report sheet.

Empty the calorimeter contents into the waste container provided. Dry the calorimeter with a paper towel.

Perform this procedure with one salt that dissolves by an exothermic process and one by an endothermic process. If time permits, run duplicate analyses of each salt investigated. Complete the calculations described in part III below as discussed in the introduction to this laboratory.

III.   Calculations

After you have completed the experiment, use your data to calculate ΔH_{soln} for each ionic compound that you studied. No specific instructions for this calculation are given in this experiment. Consult your notes, textbook, and *Experiment 8* in this laboratory manual if you need to review how to complete this calculation. Compare your calculated value of ΔH_{soln} to literature values, **but use your own data in further calculations**.

Using **your** data for the exothermic and endothermic processes that you studied, design a hot pack and cold pack. Both packs must have a total mass (salt plus water) of 400 g. The hot pack should be able to reach a temperature of 60°C starting from 23°C. The cold pack should be able to reach a temperature of 10°C starting from 23°C.

Laboratory Report
Experiment 14
Hot & Cold Packs

Name: _____

Date & Time: _____

Lab Partner: _____

I.   **Measurements and calculations of ΔH_{soln} for exothermic solution process**

Ionic compound used: _ _ _ _ _ _ _ _ _ _	Trial 1	Trial 2
Mass of dry calorimeter		
Mass of calorimeter containing H_2O		
Net mass of H_2O		
Mass ionic compound used		
Initial temperature of solution		
Final temperature of solution		

Ionic compound used: _ _ _ _ _ _ _ _ _ _	Trial 1	Trial 2
m_{soln} (total mass cmpd + H_2O)		
ΔT		
q_{soln} (in J)		
q_{soln} (in kJ)		
Molar mass compound used		
Moles compound present		
ΔH_{soln} in $kJ \cdot mol^{-1}$		
Average ΔH_{soln} in $kJ \cdot mol^{-1}$		
Accepted value ΔH_{soln} in $kJ \cdot mol^{-1}$		
Percent error		

II.   Measurements and calculations of ΔH_{soln} for endothermic solution process

Ionic compound used: _ _ _ _ _ _ _ _ _ _ _ _	Trial 1	Trial 2
Mass of dry calorimeter		
Mass of calorimeter containing H_2O		
Net mass of H_2O		
Mass ionic compound used		
Initial temperature of solution		
Final temperature of solution		

Ionic compound used: _ _ _ _ _ _ _ _ _ _ _ _	Trial 1	Trial 2
m_{soln} (total mass cmpd + H_2O)		
ΔT		
q_{soln} (in J)		
q_{soln} (in kJ)		
Molar mass compound used		
Moles compound present		
ΔH_{soln} in kJ·mol^{-1}		
Average ΔH_{soln} in kJ·mol^{-1}		
Accepted value ΔH_{soln} in kJ·mol^{-1}		
Percent error		

III. Design of hot and cold packs

A. Hot pack

_____ g of _____ and _____ g of H$_2$O

Calculations

B. Cold pack

_____ g of _____ and _____ g of H$_2$O

Calculations

General Instructions
for
Computer–Interface Systems

You can adapt some experiments in this manual so that your data are collected, stored and analyzed on a computer. These instructions introduce you to general concepts that underlie this approach to chemical experimentation. Your instructor will give you specific details of setting up individual experiments when computers are to be used.

Every experiment—even though you may never think about it—consists of four parts:

    (1)   Identification of the problem or scientific question
    (2)   Design of the experiment
    (3)   Collection of data
    (4)   Analysis of data

Experiments amenable to use of computers are ones in which collection of data involves measurements that generate electrical signals. The device generating an electrical signal is called a *probe*. A probe sends an electrical signal to a data–acquisition interface, and the interface in turn sends an impulse to a computer where it is recorded as a value in a spreadsheet. Software on the computer directs the interface to collect and send these bits of information at specified times. The figure below depicts the general arrangement of a probe, interface and computer.

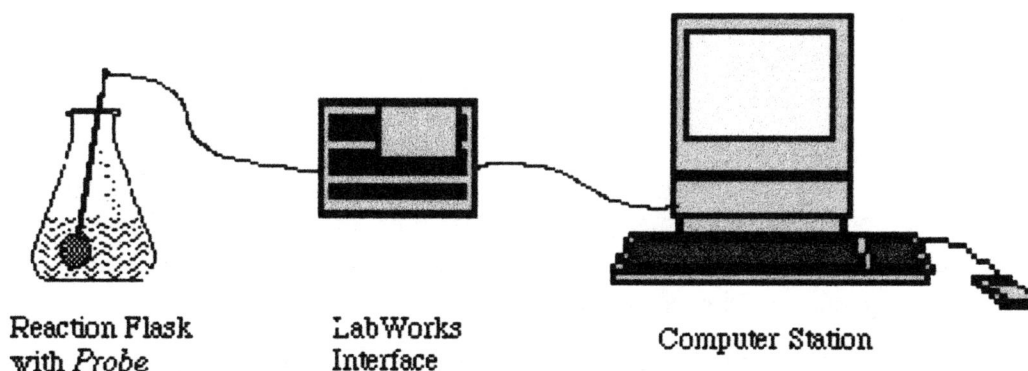

Reaction Flask   LabWorks   Computer Station
with *Probe*   Interface

Computer–assisted data collection and analysis is suitable for experiments in general chemistry that produce changes in temperature, color, acidity, pressure, or electrical conductivity. Probes are available that respond to changes in each of these properties. For example, a thermistor generates a small current that depends upon its temperature. Thus a thermistor is a probe for temperature measurements and can be used in calorimetry experiments.

In a typical calorimetry experiment, you can collect temperature measurements several times a second or every 10 minutes—whichever time you deem suitable for the reaction being studied. After the experiment is completed, your data are stored in a spreadsheet. You then use the power of the spreadsheet to perform calculations on the data and graphically represent it.

Before you begin any experiment, however, a probe must be calibrated so that its output is meaningful. You do not usually think about the calibration process in many experiments that you do in a laboratory. For example, when you use a balance, you expect that the indicated mass is accurate. At some point, however, someone must use objects of standard masses to *calibrate* the balance so that when you use the balance you will record unknown masses accurately. Without proper calibration, your mass measurements would be meaningless.

Thus you first *calibrate* the probe that you will be using with the computer–interface system. Then you *design* the experiment by selecting how often the measurement will be taken and how the data will be stored. Next you activate the system to *acquire* the data, and lastly you use the power of the computer and spreadsheets to *analyze* the collected data. Hence each experiment using the interface and computer consists of four key steps—*calibrate, design, acquire, analyze.*

LabWorksII software has been currently installed on the laboratory computers that use the Microsoft Windows operating system. When the program is accessed by double clicking on the LabWorks icon, a **Command Bar** with the buttons for these four key steps appears as shown below.

$$\left(\text{Calibrate}\right) \quad \left(\text{Design}\right) \quad \left(\text{Acquire}\right) \quad \left(\text{Analyze}\right)$$

Specific instructions are given when each button is selected in turn. The whole process from calibration to analysis usually requires less time than experiments not using the interfaces and computers, and you can often repeat these experiments several times to improve the accuracy and precision of your work.

Detailed instructions for experimental design and data analysis depend on the specific needs of an experiment and may change as instruments, interfaces and software improve over the years. Thus step–by–step instructions as to which button to push or icon to select are not included in this manual but will be provided by the instructor for each experiment adapted to use with interfaces and computers. Also you will be encouraged to utilize the power of this type of experimentation on your own.

Curve Fitting
for
Linear Relationships

Many phenomena in science and engineering vary linearly; one quantity changes in direct proportion to a second quantity. An example of this linearity is the relationship between absorption of monochromatic light and concentration of the absorbing species. This relationship is known as Beer's law, and you use it in *Experiments 9* and *10* in this manual.

Beer's law is expressed mathematically in the equation below in which *A* is the absorbance; *c*, the concentration of the absorbing species; *b*, the pathlength; and *a*, the proportionality constant called the absorptivity.

$$A = a{\cdot}b{\cdot}c$$

Usually the pathlength is held constant, and the product *ab* acts as a new constant, *k*. Beer's law then becomes

$$A = k{\cdot}c$$

Carefully prepared solutions of known concentration of a light–absorbing species are the *independent variables*, and the values of absorbance at a wavelength of maximal absorption become the *dependent variables*. A graph of *A* on the *y*–axis versus *c* on the *x*–axis yields data points that **theoretically** fall along a straight line with a slope *k* and a *y*–intercept at 0.

Because all measurements exhibit a degree of uncertainty, however, in practice experimental data points rarely all graph in a straight line. A statistical treatment of experimental values called *linear regression* or *least–squares analysis* of the data points enables you to draw the *best straight line* through the data points.

Graphing calculators and spreadsheet programs compute linear regression analysis of your data points and give you values for the slope and intercept, and you are encouraged to learn how to do these operations on your calculator and a computer. Also a simple variation of linear regression analysis, given below, enables you to calculate the best straight line through data points with just a basic calculator to speed up the mathematical manipulations. This calculation enables you to determine the slope with the theoretical line going through the origin as the theory of Beer's law indicates.

For an equation such as $A = k{\cdot}c$, with a set of values for *A* and *c*, the slope *k* of the best straight line going through the origin is given below with the symbol Σ representing "the sum of." An example is worked out on the following page.

$$k = \frac{\Sigma\,(A{\cdot}c)}{\Sigma(c^2)}$$

The table below lists typical data for the absorbance A of an *analyte* (the species that absorbs light) at four different concentrations c. Two additional columns give the product of $A \cdot c$ and c^2. Summation of the latter two columns are shown as $\Sigma(A \cdot c)$ and $\Sigma(c^2)$. The slope is calculated by the division shown.

Determination of Slope from Absorption vs Concentration Data

c (Conc. of Analyte)	A (at λ_{max})	A·c	c^2
0	0.000	0	0
0.85×10^{-4} M	0.235	0.1998×10^{-4}	0.723×10^{-8}
1.70×10^{-4} M	0.348	0.5916×10^{-4}	2.890×10^{-8}
2.55×10^{-4} M	0.624	1.5912×10^{-4}	6.503×10^{-8}
3.40×10^{-4} M	0.725	2.4650×10^{-4}	11.560×10^{-8}
$\Sigma(A \cdot c) = 4.8476 \times 10^{-4}$			
$\Sigma(c^2) = 2.168 \times 10^{-7}$			

$$\text{Slope} = k = \frac{\Sigma(A \cdot c)}{\Sigma(c^2)} = \frac{4.8476 \times 10^{-4}}{2.168 \times 10^{-7}} = 2240 \text{ M}^{-1}$$

The figure below shows a plot of these data with the *linear regression* or *best straight line* drawn based on the calculated slope. To draw the line, pick out a value of x that is easily read from the graph. Multiply this x value by the slope to obtain a value of y at that value of x. On this graph if you select an x value of 3×10^{-4} M, the calculated y value would be 0.672. Then draw a line between the origin (0,0) and this calculated (x,y) point on the graph.

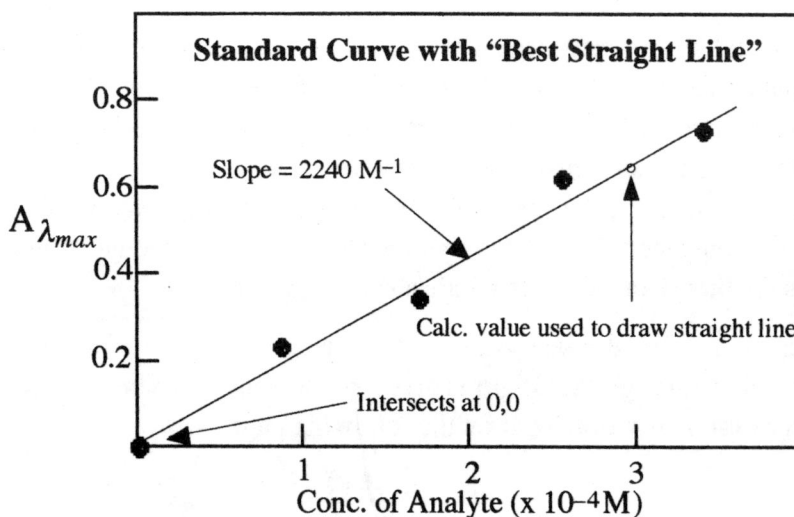

Standard Curve with "Best Straight Line"

Slope = 2240 M^{-1}

Calc. value used to draw straight line

Intersects at 0,0

$A_{\lambda_{max}}$

Conc. of Analyte (x 10^{-4}M)